T0130598

YOU

You

A NATURAL HISTORY

William B. Irvine

OXFORD
UNIVERSITY PRESS

Oxford University Press is a department of the University of Oxford. It furthers the University's objective of excellence in research, scholarship, and education by publishing worldwide. Oxford is a registered trade mark of Oxford University Press in the UK and certain other countries.

Published in the United States of America by Oxford University Press
198 Madison Avenue, New York, NY 10016, United States of America.

Library of Congress Cataloging-in-Publication Data
Names: Irvine, William Braxton, 1952– author.
Title: You : a natural history / William B. Irvine.
Description: New York, NY : Oxford University Press, 2018. |
Includes bibliographical references and index.
Identifiers: LCCN 2018012957 | ISBN 9780190869199
Subjects: LCSH: Human biology. | Human beings—Constitution.
Classification: LCC QP34.5 .I78 2018 | DDC 612—dc23
LC record available at https://lccn.loc.gov/2018012957

9 8 7 6 5 4 3 2

Printed by Sheridan Books, Inc., United States of America

To Jamie,
For helping give the cellular me
a second chance at life,
and for so much more

Contents

Acknowledgments

THE FOUNDATION FOR THIS BOOK was laid when George W. Bush was president, the research and writing were done during the Obama administration, and the final revisions were made during the Trump administration. During this period, many people and institutions played a role in making this book possible. I would like to take this opportunity to thank them.

I want to give particular thanks to Wright State University for the course reductions it granted me in the 2015-16 and 2016-17 school years. The time I thereby gained was spent writing and revising the words that follow.

I want to thank the many people who looked at and commented on book chapters, including Kevin de Queiroz at the Smithsonian Institution and his biologist brother Alan; Jayme Dyer at the Massachusetts Institute of Technology; Matthew Gale in Yellow Springs, Ohio; Erik Hill at Campbell University; Steven Karafit at the University of Central Arkansas; Massimo Pigliucci at City University of New York; Robert Riordan at Wright State University; and Sarah Roe at Southern Connecticut State University.

I would also like to thank William M. Irvine at the University of Massachusetts at Amherst. In case you are wondering, yes he is a relative of the author—as is every other person on the planet—but not a close relative.

I would like to thank the anonymous readers at Oxford University Press for their helpful comments on my proposal and offer a special thanks to the anonymous reader who read large parts of the book.

A special thanks goes to Jeremy Lewis, science editor at Oxford University Press.

And finally, I want to thank Woods Hole Oceanographic Institute for permission to use their illustration of all the earth's water drawn into one big drop (in Figure 16.3), and to David M. Hillis, Derrick Zwickl, and Robin Gutell for permission to use their innovative trees of life (in Figures 4.3 and 4.4).

Introduction

YOUR MULTIPLE IDENTITIES

YOU ARE MANY THINGS. You are, first and foremost, a person. You may or may not be a parent, but you are necessarily someone's son or daughter. If pressed for more detail, you might go on to identify yourself as being, say, an accountant, a recreational bassoonist, and a member of the middle class. You might add that you are a citizen of a certain country and that you are of a certain ethnicity and race.

Ask a scientist what you are, though, and you will get a radically different description. An evolutionary biologist, for example, might see you primarily as a member of the species *Homo sapiens*, whereas a microbiologist might tell you that what you are, essentially, is a collection of cells. She might add that for you to remain healthy, you must share your body with trillions of microbes. These single-celled organisms live in your gut and various other organs, and they inhabit every square millimeter of your skin. There are so many of them that if we did a census of the cells in and on your body, we would find that nine out of ten of them not only aren't human, but they belong to a different domain of living things than your human cells do.[1] And even stranger, it turns out that your human cells have within them the descendants of ancient bacteria. These *mitochondria*, as microbiologists call them, have their own distinct DNA, and without the power they provide, you would not be the magnificent multicellular organism that you are.

A physicist will likely take issue with the microbiologist's characterization of you. Yes, you are a collection of cells, but this doesn't get to the bottom of things, since those cells are themselves collections of atoms. The atoms in question came into existence long before you did; indeed, many of the hydrogen atoms within you have been around since a few minutes after the Big Bang, nearly 14 billion years ago. The carbon, oxygen, and nitrogen atoms in you were at one time hydrogen and helium atoms that subsequently fused into heavier atoms in the core of a star. And the only way they could ultimately become part of you is for that star to eject them in the hyper-violent event known as a *supernova*.

This, however, is only one of the adventures your atoms experienced before joining you. It is entirely possible, for example, that some of your hydrogen atoms were, not long ago, components of gasoline molecules. Furthermore, many of your nitrogen atoms, before joining you, had been struck by lightning or spent time in the reaction vessel of a fertilizer plant. And most of your atoms were previously part of another living thing—a corn plant, perhaps, or the cow that ate that corn. Some of them might even have previously belonged to another person.

Although your atoms are themselves ancient, most of them have been with you for only a small fraction of your life: it has been estimated that in the course of a year, 98 percent of your atoms are replaced by other atoms. Likewise, your cells are constantly dividing and dying, meaning that the average cell is only perhaps a decade old. Your atomic identity and cellular identity, then, are in a constant state of flux, and although your driver's license might imply that you are ready for retirement, your cells are, on average, mere children in terms of the amount of time they have spent with you, and your atoms are the equivalent of toddlers.

Unlike your cells, your atoms have nothing to gain from being part of you. If they could think, they would probably feel about their connection with you the way the passengers in a transatlantic jumbo jet feel about each other: "Yes, we are all thrown together for a time, but when the flight is over, we will be going our separate ways." And rest assured that after your death, long after your cells have putrefied, your atoms will continue their journey.

It is very likely that within a few decades of your death—and maybe much sooner, depending on what is done with your remains—your atoms will have rejoined the outside world, where they will experience new adventures. Furthermore, just as your atoms might have been part of another person before they joined you, they might become part of another person after you die. In this way, you will experience an afterlife, although not necessarily the one you had imagined.

You probably wouldn't hesitate to identify yourself as an independently living organism, but an ecologist would challenge this "independence." He would point out that you could not have evolved in the absence of other living things. Likewise, your continued existence depends on the existence of other organisms. Without them, what would you eat? And even if you came up with a way to manufacture food from chemicals, you would be in big trouble without gut bacteria to help you digest it. Also, if the earth weren't covered with plants and the ocean weren't full of phytoplankton, you would, after a time, run out of oxygen to breathe.

A geobiologist, on hearing this, might add that it is also a mistake to treat Earth and the life that inhabits it as independent entities. Our planet has obviously shaped life, but it has in turn been shaped by that life. Had life not evolved, two thirds of the minerals in its crust would not exist, and its oceans would have a radically different chemistry. Consequently, a case can be made that rather than thinking of Earth as a planet that harbors life, we should think of it as a living planet.

And finally, a geneticist, on hearing these discussions, might chime in that if you really want to understand what you are and how you came to be here, you need to learn about your genes. From the genetic point of view, you are being thoroughly exploited. Your genes have cleverly wired you so that you have the survival instinct and sex drive that allow them to propagate. Why do fools fall in love? Because their puppet-master genes have pulled the right strings! Your genes also propagate by making you want to help people who share your genes—most notably, your children. Because of their ability to exploit the organisms in which they reside, genes are the longest-enduring biological entity on our planet. Some of yours have been around for more than a billion years and can be found in trillions of other organisms.

THIS IS A BOOK ABOUT SCIENCE. In it, I describe discoveries made in physics, astronomy, biology, geology, and other disciplines. The science in question, though, can best be described as humanistic: the pages that follow are written with non-scientists in mind, and my goal in writing them is for readers not only to understand how, according to science, they and their world came to exist, but to take this science personally. My goal is also to reveal to readers their multifaceted identity—yes, you are a person, but you are also a member of a species, a collection of cells, a collection of atoms, a gene-replication machine, and a component of the living planet we call Earth.

This is also a book about trees, and by this I have in mind not the kind we find growing in forests but the kind used in diagrams that demonstrate the relationships between things. You have a family tree, of course. It shows how you are related to other people. Your species, however, also has a "family tree"—better known as *the tree of life*—that shows how it is related to other species. Your cells have cellular family trees, showing which cell was their "mother." Your genes have genetic trees, which might or might not track along your family tree. The sun that warms you even has a tree, since it owes its existence to one or more "parent stars" that exploded at least 4.5 billion years ago. (And spoiler alert: there is reason to think that the sun has a long-lost brother star.)

At this point, a concern might arise: perhaps I am relying too heavily on science in my explanation of what you are and how you came to be. Science is great at telling us how material objects, including your body, came to exist. But what if you are essentially an immaterial entity—a mind or maybe a soul—that just happens, for the time being, to inhabit a body? Since I am a philosopher, I am happy to consider this possibility and do so in the closing chapters of this book. I also address some of the questions that arise regarding the meaning of life.

As the result of learning more about who and what you are, and about how you came to be here, you will likely see the world around you with fresh eyes. You will also become aware of all the one-off events that had to take place for your existence to be possible: stars had to explode, the earth had to be hit 4.5 billion years ago by a planet and 66 million years ago by an asteroid, microbes had to engulf microbes, the African savanna had to undergo climate change, and of course, any number of your direct ancestors had to meet and mate.

It is difficult, on becoming aware of just how contingent your own existence is, not to feel very lucky to be part of our universe. This, at any rate, is the effect that doing the research for this book has had on me.

PART I

Your Deep Ancestry

1

Your People

FAMILY TREES MAKE VISIBLE our relations to others. We care about these relations because they determine, to a considerable extent, how we behave toward people. We behave differently toward a brother than toward a cousin, and we behave differently toward our own children than toward our nephews and nieces. As a result, even if we have never met a relative, we will likely feel a special connection with him or her.

Suppose family-tree research leads us to an ancestor who was a soldier in the Revolutionary War or a slave on a ship. We might wonder what it was like to live that life in that world. This in turn might trigger in us an interest in history that we hadn't previously had. And if we are a contemplative individual, it will occur to us that our own descendants might someday, when they learn of *our* existence, struggle to imagine what it would be like to live our life in our world. They also might pity us for all the things we lacked that they take to be essential for human happiness. Or—who knows?—they might instead look back at us with envy for all the things we had that were subsequently lost.

Another thing that motivates people to do ancestral research is the hope that they will discover an illustrious ancestor, maybe a famous artist, politician, or scientist. Or better still, they might uncover evidence that "blue blood" runs through their veins. Before we get too excited about finding kings and emperors in our family tree, though, we need to keep in mind that such discoveries are much more common than one might think. This is because rulers often use their power to have lots of sex, which in the past, in the time before reliable contraception, would have meant siring lots of offspring, both legitimate and illegitimate. Genghis Kahn, for example, is thought to have fathered hundreds of children, and his sons apparently followed the example set by their old man. Other ancient rulers seem to have been similarly inclined.[1] As a result, as you go back in your family tree and encounter ever more direct ancestors, you should not be surprised to find a king or emperor lurking there. What would be surprising is if you failed to find one.

People are generally excited to find someone like Confucius in their family tree. They might take this as evidence that they carry the "Confucius gene" and are therefore likely to be wiser than most people. Such behavior seems inconsistent, though, since these same individuals might take pride in how different they are from their own parents, only one generation back. It is also curious how on finding, say, Genghis Khan in their ancestry, they don't worry that inheriting the "Genghis Khan gene" will make them prone to cruelty. Genes, they will tell themselves, don't work that way.

In doing ancestral research, we might also encounter unpleasant surprises. In examining our father's military records, for example, we might discover that because of his blood type, he could not be our genetic father. This might launch us on a quest to find our "real" father. It is also possible for us to discover that the woman who raised us from birth isn't our biological mother. This will be the case if the woman who raised us stole us. It will also be the case if she adopted us but chose to keep the adoption a secret, or if there was a baby mix-up at the hospital. Again, the realization that we have never met our biological mother is likely to kindle within us a desire to do so.

Even if we are confident that the woman who raised us is our birth mother, she might not be our genetic mother. Thanks to medical breakthroughs, it is possible for doctors to take an egg from Ms. A, fertilize it with sperm from Mr. B, implant it in the womb of Ms. C, and when it is born, turn it over to adoptive parents Mr. D and Ms. E. In this case, Ms. E will be the child's legal mother, Ms. C will be the child's birth mother, and Ms. A will be the child's genetic mother.

And things can be even more complicated than this. In 2014, Emelie Eriksson gave birth to Albin, a son. The birth was unusual inasmuch as Albin had been carried in a transplanted womb, meaning that although Eriksson was Albin's genetic mother and "vaginal mother," she was not his "uterine mother." And the story has another wrinkle: Eriksson had obtained the womb from her mother, Marie. Consequently, Marie was both the child's genetic grandmother and uterine mother. Pity the poor genealogist who attempts to construct a family tree that displays all these relationships.

A DETAILED FAMILY TREE will include all the children a couple had or adopted, as well as any remarriages that might have taken place. A *basic* family tree, though, will be concerned only with a person's direct genetic ancestors. Your own tree will start with you. Above you will be your (genetic) father and mother, above them will be their fathers and mothers, and so on. The logic of a basic family tree is therefore straightforward: for each person listed on the tree, there must be two entries above it, for the one male and one female who are that person's genetic parents. (Yes, there can be questions about who someone's genetic parents really are, but for present purposes, we can ignore this fact.) See the top half of Figure 1.1 for the basic family tree of President John F. Kennedy.

Besides constructing family trees, we can construct trees of descendants (which are sometimes confusingly referred to as *family trees*). To construct a tree of descendants, we start by drawing a box to indicate the person whose tree it is. Below it will be boxes depicting that person's (genetic) offspring, below them will be boxes depicting *their* offspring, and so on. In making such a tree, we don't list the people with whom a person had offspring. For example, if a woman had three children by three different men, all three children will appear on her tree, but the men who fathered them won't. When these men construct their own trees of descendants, though, each will list the child he had with this woman, together with any other children he might have fathered. For an example of a tree of descendants, see the bottom half of Figure 1.1.

Notice that in John F. Kennedy's tree of descendants, the boxes belonging to John Fitzgerald Kennedy Jr. and Patrick Bouvier Kennedy have no branches below them. This is because they died without having any offspring of their own. These *terminal branches*, as they might be called, are striking things if one takes the long view regarding life on Earth. An individual who dies without having reproduced is doing something that *none of his direct ancestors, going back to the first living organism on the earth, did.* In saying this, I'm not suggesting that we have a duty to reproduce; indeed, I would argue quite the opposite. I'm simply pointing out that a person who fails to reproduce is breaking a very long reproductive chain.

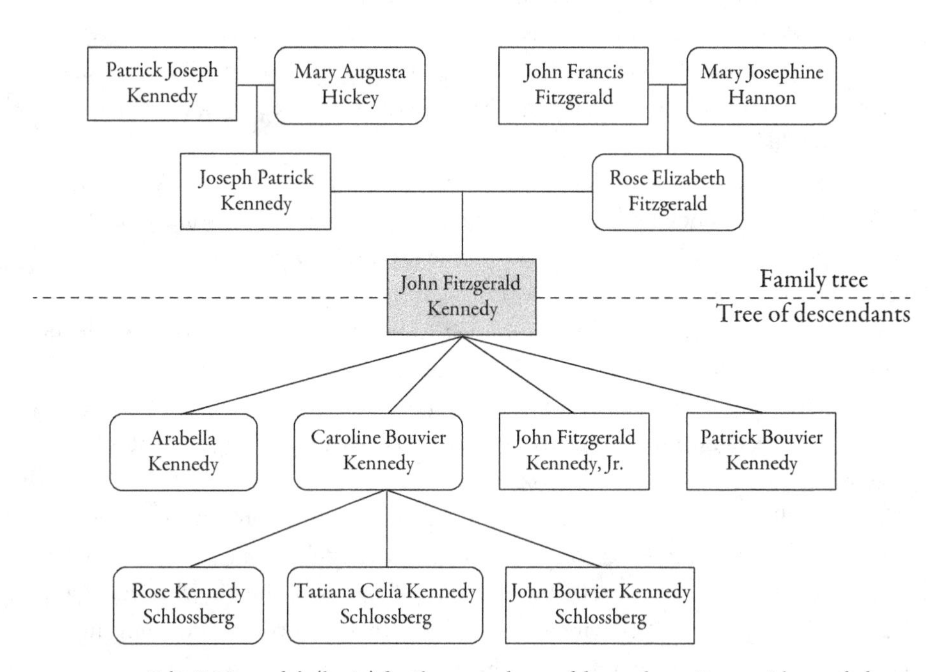

FIGURE 1.1. John F. Kennedy's (basic) family tree and tree of descendants. Boxes with rounded corners indicate females.

The logic of trees of descendants is significantly different from that of family trees. In a family tree, each entry will have exactly two entries above it, for the two parents. In a tree of descendants, though, an entry can have any number of entries below it. My tree of descendants, for example, has two entries below it, since I have two children. My wife's tree of descendants looks the same as mine, since we had all our children together. By way of contrast, the tree of descendants of Nadya Suleman, the "Octomom," has fourteen entries below it—eight for the octuplets she had, and six more for the children she had before having those octuplets. Consequently, although any two basic family trees will look structurally alike, two trees of descendants can be radically different in appearance. And as the Octomom example demonstrates, although the growth of your family tree, as you go back in time, will be impressive, the growth of your tree of descendants, as you go forward in time, can be astonishing.

A FEW YEARS AGO, I USED an internet company to do ancestral research. In the process of doing so, I made several discoveries. I learned, for example, that in 1940, according to the census of that year, my mother worked 48 hours per week as a hospital receptionist and had an annual income of $1,427. This works out to 60 cents an hour—about $10 an hour in today's dollars. I also learned that her father had worked as a "piler" in a sawmill. I was puzzled by the meaning of the word *piler* but finally concluded that my grandfather had spent his days piling freshly sawed lumber into neat stacks, a job that I would have hated. Indeed, my worst day at work is probably better than his best day—something I like to remind myself of when I am in the middle of grading a stack of student exams. I also discovered that this lumber piler had been the eighth of ten children. His family would presumably have been even larger had his father not died while my grandfather was still a child.

What I came to realize, in doing this research, is that the further back I went, the harder it became to obtain reliable information. Records might be missing or might not have existed to begin with, and when records did exist, they were often handwritten and therefore semi-legible. And to top things off, as I went back, the spelling of names started changing. At one point in my research, I had traced my ancestry back to Welsh coal miners. It was a discovery that delighted me—surely it indicated that I had a capacity for hard work, right? But alas, it turned out that my ancestral research had been derailed by a change in the spelling of my surname from *Irvine* to *Irvin*. Such changes are hurdles that genealogists have to clear if they are to correctly construct a family tree.

Spelling changes can be more dramatic than this and therefore much harder for a genealogist to figure out. Furthermore, if you go far enough back in your ancestral research, you will come to the time before surnames. We can get an idea of what things were like back then by taking a look at the *Domesday Book*, which contains the information obtained in the great survey of England that was conducted in 1085–6. This survey was ordered by William the Conqueror, who wanted to find out who in England owned what and how much it was worth, so he could tax his new subjects. Among the information gained in the survey were the names of the people connected with the various estates, but because people lacked surnames, listing them proved challenging.

There were, for example, lots of surnameless *Roberts*. To distinguish between them, the Domesday commissioners listed them by telling where they lived (*Robert of Dun*), their occupation (*Robert the hunter* and *Robert the interpreter*), or their appearance (*Robert the bald*). *Robert the bastard* was listed by the circumstances of his birth, and *Robert the lascivious* was listed by what was apparently an unfortunate character trait. (How would you like to be listed in official records and be remembered a thousand years later as *Robert the lascivious*?) There were so many *Roberts* that one commissioner, apparently at his wits' end trying to come up with descriptive words, listed one of them simply as *another Robert*.[2] Good luck trying to trace your family tree through this period!

And besides seemingly popping into existence out of nowhere, surnames can go extinct. This is what happens when the last person who uses a name dies. Among the surnames that have apparently gone extinct (in Britain, at least) are such charmers as *Miracle, Relish, Bread, Puscat*, and *Birdwhistle*—also spelled *Birdwistle, Birdwhistell, Birtwhistle*, and *Burtwhistle*.[3]

IN MOST HUMAN ENDEAVORS, THE LONGER you work at a task, the closer you are to completion. This, however, is not the case with ancestral research: the further back you go, the more work there is to do. In doing the research for your two parents, you turn up four grandparents who need to be researched, and in doing the research for them, you turn up eight great-grandparents who need to be researched. As a result, doing ancestral research is like fighting a mythical Hydra: for each head you chop off, two new heads grow.

Suppose, though, that in doing your ancestral research, you vow not to become dispirited: you are going to construct your family tree all the way back to 1 AD. Suppose, too, that you find a way to overcome all the changes in surname spelling, all the missing and inaccurate records, and so on. If we assume that 25 is the mean age at which women have children,[4] your research would take you back 80 generations. At the topmost level of your tree, there would be 2^{80} spaces to fill in—or, stated in more conventional terms, 1.2×10^{24} spaces. (By way of explanation, 10^{24} can also be represented by a 1 with 24 zeroes after it: 1,000,000,000,000,000,000,000,000.) And to get to this topmost level, you would have had to fill in the 79 levels below it—a Herculean task, if ever there was one.

To put these numbers into perspective, suppose you insisted on displaying your family tree on a piece of paper and allowed a rather stingy 1 square centimeter (0.4 inch) of space for each person listed. You would need a piece of paper a bit more than 80 centimeters (32 inches) high, which is not so bad. But the piece of paper would have to be very wide—more than a billion times wider than the solar system!—in order for you to show its top line.[5]

Suppose that on realizing this, you decided instead to store your family tree information in a computer. You would be hard pressed to find one with enough memory. The computer on which I am writing this book has a terabyte—1 trillion bytes—of memory. If you allowed a stingy 10 bytes per entry by making use of some kind of compression algorithm, the top line of your family tree back to 1 AD would require 10 trillion of my computers to hold its data.[6] And remember, this is just the memory requirement for the top line; there would be 79 previous lines, the information of which also needed to be stored, and doing so would require another 10 trillion computers like mine.

It is easy for someone who embarks on a program of ancestral research to experience a growing sense of hopelessness. In my own research, I went back only a few generations before deciding that I had more pressing things to do.

PEOPLE TEND TO BE PROUD of their ethnicity. Ask them where "their people" are from, and they might tell you China or Spain. Or they might tell you that they are a mix of French, Irish, and Cherokee Indian. Politicians are particularly likely to boast of belonging to a certain ethnic group, in part because they think it will influence how people vote. Voters might assume, for example, that a candidate who is one-sixteenth Cherokee will take a special interest in issues that affect American Indians—and given the way we feel about our ancestors, it is an assumption that could very well be true.

In some cases, people have done the research necessary to trace their ancestry back to a particular region or ethnic group. In many more, they will rely on information provided by a relative, who in turn got it from a relative. This information, although trusted, won't necessarily be correct. In yet other cases, people will infer their ancestry simply by looking in a mirror. They might take their high cheekbones as evidence of American Indians in their ancestry, or they might take blond hair and blue eyes as evidence of a Nordic ancestry.

Sometimes people determine their ancestry not by trying to construct their family tree but by contemplating their surname. If it is *Lopez*, for example, it seems safe to conclude that they are ultimately of Spanish ancestry, and if it is *Donati* that they are of Italian ancestry. The problem is that in cultures in which children take their father's surname, with each generation back you go, the patrilineal ancestor responsible for that surname will represent an ever-diminishing portion of your ancestry. Go back 10 generations, and you might have only one ancestor who is a Lopez (or a Donati), with the other 1,023 direct ancestors having, say, solidly Germanic surnames—Müller, Schmidt, Fischer, and so on. Bottom line: your surname is not a reliable indicator of your ancestry.

Sometimes it is possible to establish your ancestry not by means of ancestral research but by means of genetic tests. In some parts of the world, people were so rooted to a region that they extensively interbred and as a result developed unique genetic "fingerprints."

In the nineteenth and twentieth centuries, they might finally have left that region, but they took with them region-specific genes that allow their descendants, by taking a DNA test, to trace their origins to, say, Sardinia. But such tests, as we are about to see, can be misleading.

WHEN MY SON WAS MAYBE SEVEN YEARS OLD, I asked him if he knew where milk comes from. "Of course!" he answered. "The supermarket." This answer was indisputably correct; it was nevertheless unsatisfactory, inasmuch as it didn't get to the bottom of things.

This same phenomenon can be observed when you ask people about their ethnic roots. "My people," they might say, "come from Sardinia, and I have a DNA test that proves it." What such people are doing is picking an arbitrary stopping point in their ancestral research and declaring *that* to be their point of ancestral origin, but of course it isn't, since such claims give rise to a rather obvious follow-up question: "You say your ancestors came from Sardinia. Some of them doubtless did. It might even be the case that there was a period during which *all* of your direct ancestors who were then alive lived in Sardinia. But where did *their* ancestors come from?" It is a question, I have found, that tends to bring conversations to a halt. You watch as the person who made the ethnic claim tries to defend it. He might want to say that the ancestors of his Sardinian ancestors *also* came from Sardinia, but he will realize that for this to happen, life would have to have originated on Sardinia, a claim that even the most ardent Sardinian chauvinist will be reluctant to assert.

Along similar lines, if you boast of your British ancestry, you haven't gotten to the bottom of things, inasmuch as your British ancestors came from outside of Britain. Romans invaded Britain in the first century, Germanic Anglo-Saxon tribes migrated there in the fifth, Vikings plundered it in the ninth, and Normans conquered it in the eleventh. And even your Celtic ancestors who were there before the Roman invasion came from somewhere else.

So where will your search for your ultimate ethnic origins lead you? According to paleoanthropologists, our species, *Homo sapiens*, arose about 200,000 years ago[7] in and around the Rift Valley, which runs through what is now Ethiopia, Kenya, and Tanzania. It was only 60,000 to 70,000 years ago that they left Africa.[8] Many took the northern route out, through Egypt and into the Middle East. Others probably headed east, across the Bab-el-Mandab Strait and entered what is now Yemen, at the southern tip of the Arabian Peninsula (see Figure 1.2). This strait is now 20 miles wide, but 70,000 years ago, the earth was in the midst of the ice age from which we are still emerging. Sea level would therefore have been much lower, meaning that the strait would have been much narrower. And because of the cooler climate of that time, places like Saudi Arabia and Iraq, which are now covered mostly with inhospitable deserts, would have been much wetter and therefore would have had rivers and lakes, along with abundant vegetation.

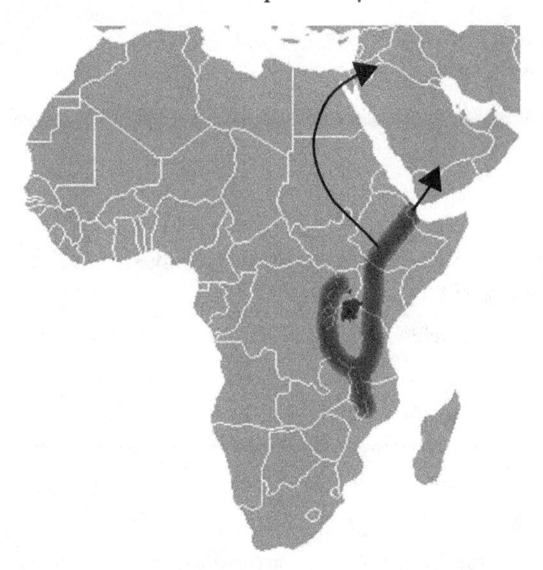

FIGURE 1.2. The shaded area is the Great Rift Valley of Africa, where "your people" came from. Arrows indicate the principal routes out of Africa.

It is thought that after crossing the Arabian Peninsula, many of our ancestors hugged the coast, moving across southern Iran and India, down through Indonesia, and then into Australia, which they reached 45,000 years ago.[9] One might think it natural for them also to head into Europe and northern Asia, but because of the ice age, these regions had climates that our ancestors, who had evolved in the tropics, would have found challenging.

We should not make the mistake of thinking of our ancestors who left Africa the way we think of Christopher Columbus—as brave explorers who set out in search of a new world. They might simply have wandered out of Africa as part of their daily search for food. They might also have left Africa against their will, perhaps to evade enemies. They might even have been blown out of Africa by storms that struck when they were out on rafts. In any of these cases, they would have been oblivious to the historical significance of what they were doing.

On leaving Africa, our *Homo sapiens* ancestors would have encountered archaic humans,[10] including Neanderthals. Like us, the Neanderthals can trace their ancestry back to Africa, but their ancestors had left it hundreds of thousands of years before our ancestors did. One can only wonder what encounters between our ancestors and Neanderthals would have been like. Neanderthals would have looked like us but at the same time would clearly have been different from us. How, our ancestors would have wondered, should we relate to these beings? Should we befriend them or exploit them? Should we mate with them or kill them? Or maybe mate with their women and kill their men?

It is conceivable that our ancestors experienced feelings that are rather like those that contemporary humans might experience on visiting the chimpanzee exhibit at a zoo. We look into their eyes and see that they are looking back at us. As we size them up, we realize that they are sizing us up as well. We conclude that someone is definitely "at home," so to speak, but who or what? These thoughts in turn give rise to questions about how we should relate to these beings. Is it appropriate for us to put them on display in zoos? Would we ourselves like to be similarly displayed? That is when, in many cases, we bring our internal debate to an end by turning our back on these apes and heading off to less awkward encounters—with newts, perhaps, or maybe cute little penguins.

IF YOU TRACE YOUR FAMILY TREE BACK, you will find that your ancestors lived in many different places. You might also find that many generations of them lived for an extended period in one particular place. But if you keep tracing, you will ultimately be led to Africa. According to scientists, this is where, 70,000 years ago, *all* your direct ancestors lived. Therefore, regardless of whether you self-identify as being of, say, French, Chinese, or Cherokee Indian ancestry, you are ultimately of African ancestry.

On hearing this claim, readers might try to turn the tables on me. Yes, our ancestors may all, at one time, have lived in the Rift Valley, but where did *their* ancestors come from? Surely they didn't just pop into existence. Touché!

This point is well taken. But realize that my claim is that if we trace our ancestry *as members of the species Homo sapiens*, it will lead us back to Africa, since this is where our species apparently evolved. I am perfectly willing to admit that our species itself has ancestral species, an idea I will examine in detail when I explore the "tree of life" in chapter 4.

Before I went to college, I would have described myself as being of Scottish-English ancestry. This is what my parents told me and what my surname suggested. Then, as a result of various influences, I moved my arbitrarily chosen stopping point *forward* in time and started identifying myself as an American. It is, after all, where my immediate ancestors had chosen to live. As a result of doing the research for this book, though, I have started identifying myself as a Rift Valleyan, in hopes that those who hear me ask what I mean by this. If they do, I am more than happy to fill them in on this part of their extended personal history.

To be sure, your ethnicity involves more than where on Earth your ancestors came from. Regardless of where that might have been, you can identify yourself as being, say, ethnically Jewish or ethnically Muslim. Your ethnicity provides you with an organizing principle for your life, which will affect what you eat, what you do on weekends, how you are buried, whom you marry, and what music is played at your wedding. It is far easier to adopt and live in accordance with such a principle than to have to extemporize your way through life.

In many cases, though, people don't so much adopt an ethnicity as inherit one. Consider, for example, an ethnically Jewish couple. The wife might not have been ethnically Jewish before marrying her husband; for her, the adoption of Jewish ethnicity was a deliberate act. Her husband, by way of contrast, might not have consciously chosen to be ethnically Jewish; he might instead have been raised that way by parents who were raised that way by parents who in turn were raised that way. But significantly, if this husband goes far enough back in the investigation of his ethnicity, he will encounter ancestors who abandoned their parents' religion in order to adopt Judaism, the way his wife did. Suppose, in particular, that he can trace his ancestry all the way back to Abraham. In doing so, he will have arrived at an ancestor who, in order to do what he took to be God's bidding, abandoned his country, his kindred, and his father's house.[11]

The same is true, of course, of those who are ethnically Muslim. It is impossible for *all* your ancestors to have been Muslim. In particular, *none* of your direct ancestors in the year 550 AD could have been Muslim, for the simple reason that Mohammed hadn't yet been born. Furthermore, even if you can trace your ancestry back to Mohammed himself, realize that by deliberately choosing to do what he believed to be the bidding of God, Mohammed abandoned what previously would have been his ethnicity.

Sometimes people feel bound by the ethnic choices their ancestors made and argue that it would be a major betrayal for them to abandon their inherited ethnicity. This line of reasoning has always struck me as inconsistent. If these individuals trace back their ancestry, they will inevitably encounter an *ethnic defector*—an individual who turned his back on his ethnicity to adopt a new ethnicity that was transmitted (by upbringing rather than genes) through subsequent generations. Notice, though, that if this ancestral defector had followed the injunction to remain true to the ethnicity of his ancestors, he wouldn't have switched his original ethnicity the way he did, meaning that his modern descendants wouldn't have the inherited ethnicity that they feel so bound to maintain. Strange behavior, indeed.

SO MUCH FOR YOUR ETHNICITY. What about your race? If you self-identify as being white, you are in for a surprise if you do extended ancestral research. As we have seen, 70,000 years ago, all of your ancestors lived in Africa, and these ancestors almost certainly would have been dark-skinned. In fact, research suggests that as recently as 8,000 years ago, your ancestors were dark-skinned.[12] The scientific consensus is that light-skinned people evolved only when members of our species moved to higher latitudes in Europe. In that climate, it would have been advisable for them to wear clothing during much of the year. By wearing clothing, though, they would have experienced less exposure to sunlight, and the sunlight in question would have been less intense than in Africa.

Under such circumstances, dark-skinned people would be at a disadvantage, since they would tend to be vitamin D deficient. When our skin is exposed to sunlight, our bodies make vitamin D, but the melanin that darkens skin interferes with this process.

This means that light-skinned people who lived in what is now Europe would be more likely than dark-skinned people to survive and reproduce. As a result, with passing generations the skin color of Europeans became lighter and lighter. Conversely, unclothed light-skinned people would have been at an extreme disadvantage in Africa, inasmuch as they would have been susceptible to skin cancer.[13]

Many readers will accept the principle that if both your parents are black, then you are black as well. From this principle, though, it follows that *everyone is black*. If our ancestors were black when they left Africa, their offspring would also have been black. Likewise, since these offspring were black, *their* offspring would have been black, and so on, down to us. Conclusion: in racial terms, we are all black, even if we are not dark skinned.

Even now, though, we haven't gotten to the bottom of things. Once we concede that our ancestors of 70,000 years ago were all black, we will, if we are atop our game, ask the obvious follow-up question: yes, but what about *their* ancestors? To answer this question, we might take a look at our modern-day ape cousins, including chimpanzees and baboons. Most of their skin is covered with fur, but the part we can see tends to be dark, so it seems reasonable to conclude that the ancestral species we have in common with them would also have been dark skinned.

The problem with this line of reasoning is that it assumes that the skin under an ape's fur is the same color as the skin that is visible to us. This is not the case, though. Whereas human skin is uniformly pigmented, that of the other great apes is patchy. This becomes apparent when we encounter chimpanzees with alopecia, a condition that causes them to lose their fur without affecting the color of their skin. The skin previously covered by their fur might be described as grayish pink in color.

A case can be made, then, that although your ancestors of 70,000 years ago may have been dark skinned, your ancestors of 7 million years ago—the ancestors you have in common with chimpanzees[14]—were not. They would have been covered with fur, but the skin beneath that fur would probably have been light colored, the way the skin of modern chimpanzees is. Therefore, if you think your race is inherited from your ancestors, it makes perfect sense for you, regardless of whether your skin is dark or light, to identify your race as being neither black nor white but grayish pink. And probably what makes even more sense, when someone inquires about your race, is simply to reply "human."

2

You and I Are Related

ANCESTRAL RESEARCH, AS I have said, can yield surprising results. Maybe you have an ancestor who was president of the United States. Alternatively, an ancestor might have been a slave or even a slave owner. It is also possible, as college student Shannon Lanier found out, to discover all three of these in a single generation of your family tree. Historical records and DNA tests revealed that Lanier was a descendant of Thomas Jefferson, who was both a president and a slave owner. Not only that, but Lanier was a descendant of Sally Hemings, who was both Jefferson's mistress and one of his slaves.[1] There is also reason to think, by the way, that Hemings was Jefferson's wife Martha's half-sister.[2] Family trees can be curiously convoluted things.

Another surprising consequence of ancestral research can be the discovery that you are married or engaged to a relative. In one such case, a man was looking through his fiancée's family album when he came across a picture he had seen before: a copy of it was hanging in the dining room of his grandparents' house. Confused, he started asking his relatives questions, only to discover that the person in the dining room picture was his great-grandfather and therefore that he and his fiancée were related. They decided to go through with the wedding anyway, a decision, they discovered, that made them the target of many jokes, such as that the first dance at their wedding should be to the song "We Are Family."[3]

In other cases, accidentally marrying a relative has resulted not in laughter but in anguish. These cases might involve siblings unwittingly marrying each other[4] or women unwittingly marrying their fathers.[5] Such occurrences are possible in part because of laws that allow sealing of birth certificates in order to conceal the fact that someone has been adopted.

FAMILY TREES GROW AT AN EXPONENTIAL RATE. For each generation shown on a tree, there will be twice as many entries as there were for the previous generation: you have two parents, four grandparents, eight great-grandparents, and so on. Go back to the year 1 AD, and in the top line of your tree there will, as we have seen, be 1.2×10^{24} spaces to fill. This isn't just more than the 7 billion people who now live on Earth; it is 10 trillion times more than the 108 billion people who are estimated *ever to have lived*.[6] This growth rate creates problems for anyone trying to do a family tree. Go back far enough, and you will reach a point in time at which there are more spaces on your tree than there are people to fill them with. Let us refer to this problem as the *family tree paradox*.

This paradox, which might at first seem daunting, turns out to have a simple solution. We need only keep in mind that an individual can make multiple appearances on a tree. Suppose, for example, that you marry a first cousin. You and that cousin will by definition have a pair of grandparents in common. Suppose, too, that you and your cousin subsequently have a child—let's call her *Alice*. On Alice's family tree, there will be the usual two spaces for her parents, four spaces for her grandparents, and eight spaces for her great-grandparents. Since, however, Alice's parents have a pair of grandparents in common, Alice will be able to fill in the eight spaces on the great-grandparent line of her tree with only six individuals: the grandparents that Alice's parents have in common will appear twice on her tree, once as her mother's grandparents and again as her father's grandparents. The same thing will happen, although less dramatically, if you marry not a cousin but a tenth cousin, with whom you share great-times-nine grandparents.

In our solution to the family tree paradox, we also need to keep in mind that whenever a person makes multiple appearances on your tree, his or her ancestors will also make multiple appearances. This means that the number of spaces on your family tree isn't the only thing that grows at an exponential rate as we go back in time; so does the number of people who make multiple appearances on that tree. Consequently, the 1.2×10^{24} spaces on the 1 AD level of your family tree can be filled in with far fewer than 1.2×10^{24} individuals.

For a dramatic example of this phenomenon, consider the family tree a creationist might construct. Adam and Eve, he will tell us, had two sons, Abel and Cain, and later, at the age of 130, Adam had another son, Seth, followed by additional sons and daughters. Adam's sons could not have married the unrelated girl next door: inasmuch as they were the only people on Earth, there were no such girls. What they presumably did instead is marry their sisters. Subsequently, Cain and one of these sisters had Enoch. If Enoch did his family tree, it would show Cain and Cain's sister as his parents, and it would list Adam and Eve *twice*, as both his paternal grandparents and his maternal grandparents (see Figure 2.1).

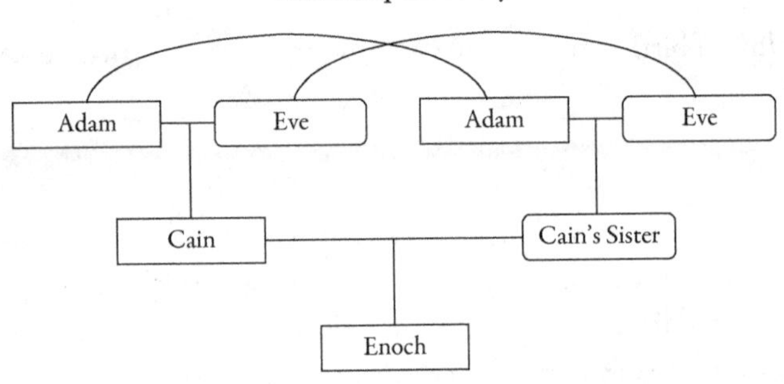

FIGURE 2.1. The family tree of Enoch, according to the Old Testament. Notice the double appearance of Adam and Eve in this tree. If the biblical story of creation were correct, this couple would likewise appear repeatedly along the top line of your family tree.

Adam and Eve's other grandchildren would have similar trees. Furthermore, the children of these grandchildren would have family trees that at the great-grandparent level would list Adam and Eve, as a couple, four times; and *their* children would in turn have trees that would list Adam and Eve, as a couple, eight times as great-great grandparents. This pattern would continue down to Adam and Eve's present-day descendants. Consequently, although a creationist might not be able to fill in most of the middle levels of his family tree—the records are missing—he will tell you with great confidence that his tree will "top out" in a level that shows Adam and Eve, as a couple, repeatedly appearing. It is, to be sure, a curious tree, but it is a necessary consequence of creationist beliefs—which beliefs, for the record, I do not share.

LOOKING AT THE FAMILY TREES of European nobility, we can find dramatic examples of this multiple-appearance phenomenon. Most people have eight distinct individuals as great-grandparents, but Alfonso XII of Spain has only four. Charles IV of Spain plays a prominent role in this same tree. He appears twice as Alfonso's great-grandfather. He also appears twice as Alfonso's great-great-grandfather: he is Alfonso's mother's mother's mother's father, as well as his father's mother's mother's father. Princess Maria Luisa of Parma also appears both as Alfonso's great-grandmother and as his great-great-grandmother.

In the past, royals married relatives because they wanted to keep wealth and power in their family. Commoners also tended to marry relatives but for an entirely different reason: they lived in a time when it was hard to leave their place of birth. Consequently, they ended up marrying someone who was born nearby and was therefore likely to be a relative.

Marriage between relatives is particularly likely on islands. Iceland's population, for example, is rather inbred, giving rise to the "bump the app before you bump in bed" campaign. Two Icelanders who have just met and are sexually attracted to each other can use this smartphone app to find out whether they are cousins.[7] In Ireland, marriage between cousins is also common. In the previous chapter, I presented John F. Kennedy's family tree. Had I shown the extended tree for his Irish ancestors, we would have seen that Kennedy's mother's parents, John Francis Fitzgerald and Mary Josephine Hannon, were second cousins, meaning that they had great-grandparents in common. And on the island of Sardinia, the population is sufficiently inbred that Sardinians have developed unique genetic "fingerprints." Indeed, researchers can, on the basis of DNA analysis, determine not only that someone came from Sardinia but make a good guess about *which Sardinian village* he came from.[8]

This last research has yielded unexpected results. In 1991, a frozen body was found in the Alps. It turned out to be 5,000 years old and was christened *Ötzi, the Iceman*. Scientists did a DNA test and discovered that Ötzi's genes bear a striking resemblance to those of modern-day Sardinians. "It's a real mystery," said geneticist Carlos Bustamante. "Did he or his ancestors travel to the Alps from Sardinia, or did the Europe of 5,000 years ago more closely resemble Sardinia and Corsica? It's a fascinating question in part because it addresses how rapidly people spread across Europe, and how far they rambled."[9]

And to deepen this mystery, a few years after Ötzi's Sardinian connection was revealed, it was discovered that a woman who lived in Ireland 5,000 years ago also had DNA that resembled that of Sardinians. One would not expect such a connection between Ireland and Sardinia. It also turns out that some of the people who settled Ireland 4,000 years ago came from Eastern Europe.[10] In other words, an Irish ancestry, like a British ancestry, turns out to be rather more complex than those who boast of such ancestries generally realize.

Yes, past nobility might have intermarried, as might people stuck in a region or on an island. But what about we moderns who are living far from our place of birth? It turns out that it isn't just likely that you are related to any spouse or lover you might have, it's a sure thing. The only question is how far back you have to go to find the intersection between your family trees. I will defend this claim in a moment, but first let us explore the consequences of inbreeding.

MOST PEOPLE ARE AVERSE to mating with close relatives. They are also averse to other people doing so. As a result, consanguineous marriages have been outlawed in many places. These laws differ, though, with regard to the forbidden degree of consanguinity. In the United States, marriage of second cousins is legal in every state, and marriage of first cousins is legal in seventeen states, and, if certain conditions are met, in seven other states as well. In Pakistan, marriage between first cousins is not only legal but common. In Ethiopia, by way of contrast, marriage between even sixth cousins—individuals who share great-times-five grandparents—is frowned upon.

If you ask people why they are averse to marriage between close relatives, they will often point to the birth defects to which such unions can give rise. Along these lines, consider the case of Charles II of Spain. His family tree, like that of Alfonso XII of Spain, is convoluted. For example, Charles's mother was her father's niece—meaning that her father married his sister's daughter. In part because of this inbreeding, Charles II was a genetic wreck. He has been described as being "short, lame, epileptic, senile, and completely bald before thirty-five."[11] He was also impotent, and his failure to produce an heir arguably triggered the War of Spanish Succession, in which perhaps a million people died.

A case can be made, though, that the dangers of inbreeding are overstated. Yes, marriage between first cousins doubles the chance that their offspring will have birth defects—but the doubling is from 2 percent to only 4 percent.[12] Infant mortality also rises to 4.4 percent, but this is the same rate as a woman over 40 will experience if she has children with someone who isn't a close relative.[13] We aren't in a panic about older women having children, so doesn't it seem inconsistent for us to be concerned about first cousins having them? Perhaps we overstate the risks because we have been indoctrinated with the "incest taboo": at the gut level, we *want* incest to have terrible consequences.

Turning to the animal world, we can find lots of cases in which inbreeding—even intensive inbreeding—didn't have genetically disastrous consequences. All thoroughbred horses, for example, can trace their ancestry to one of three stallions and to a few dozen mares. Likewise, the various breeds of dogs are the result of intensive inbreeding. As a result, problems do arise: Dalmatians, for example, are prone to deafness, and dachshunds experience back problems, but such cases are the exception rather than the rule.

To be maximally useful in experiments, laboratory mice should be as genetically similar as possible. That way, differences between experimental groups can be attributed to differences in what was experimentally done to them, not to differences in their genetic makeup. To achieve this genetic homogeneity, mouse breeders resort to extreme inbreeding. They mate two mice, producing a bunch of brothers and sisters, who in turn are mated, and so on, down through maybe 20 generations.[14] The resulting mice, which seem perfectly healthy, are not clones of each other, but they have very similar genes.

This is not to say that you can mate any two mice and end up with a successful inbreeding program. If two mice have problematic genes, those genes will be amplified by inbreeding, and the resulting mice will either die or not be healthy enough to use in experiments. But if the "founder mice" have the right genes, inbreeding isn't an issue.

Another place we find effective inbreeding is in conservation biology. The population of elephant seals, for example, was reduced to only 20 by hunting; it has subsequently recovered, thanks to inbreeding of the remaining animals. Likewise, the number of giant tortoises on the Galapagos Islands was down to 15 in the 1960s; it is now 1,500.[15]

Is it possible for a species to be down to two members and recover? Absolutely. In 2004, a hurricane hit the Caribbean, flooding several islands and thereby killing their resident lizard populations. The next year, scientists selected seven male-female pairs of brown anole lizards from a nearby island that hadn't flooded and put one pair on each of seven "sterile" islands. All seven pairs were able to repopulate their respective islands.[16] And if two lizards can repopulate an island, a single "pregnant" female should also be able to do so.[17]

These examples show that when a population is sharply reduced, in what biologists refer to as a *bottleneck event*, it can bounce back without the survivors being genetically defective. These survivors will, however, have lost their genetic diversity. As a result, if their environment changes in a dramatic fashion, there won't be any genetic outliers around who can cope with that change. A species that survives a bottleneck event will therefore subsequently run a greater risk of extinction.

AS I HAVE SAID, NO MATTER WHO you mate with, you are mating with a relative. This is because *all* people are related, in the sense that they have an ancestor in common. More generally, any two living organisms have an ancestor in common. You and a randomly chosen chimpanzee, for example, have an ancestor in common. It lived about 7 million years ago and was neither a human nor a chimpanzee; it was instead a member of a species that is no longer around, having been slowly transformed, by evolutionary processes, into three other currently existing species: humans, chimpanzees, and bonobos. You also have an ancestor in common with the millions of *E. coli* bacteria that are now roaming your intestines. That ancestor would have been a very ancient microbe.

Realize that for two currently existing organisms to be *completely* unrelated, their family trees could never intersect, and by *never* I mean all the way back to the first living organism on Earth. The only way this could happen is if life had arisen twice on Earth, with some currently existing organisms being able to trace their ancestry back to one of those first-living things and other currently existing organisms being able to trace their ancestry back to the other first-living thing. In chapter 7, I will present "smoking gun" evidence that this is not the case—that all currently existing organisms *do* have an ancestral organism in common and are therefore related.

This means that all currently existing people have an ancestor in common, raising the question of who this ancestor was. If complete and accurate genealogical records were available, identifying the common ancestor would be a fairly straightforward task. In the absence of such records, though, researchers are forced to rely on computer modeling. One such model, based on a number of demographically reasonable assumptions, suggests that some person who roamed the earth about 2,300 years ago[18] is the direct ancestor of everyone now alive. They refer to this individual as the *Most Recent Common Ancestor*—the *MRCA*, for short.

That such an ancestor would exist might seem implausible. To be the MRCA, a person would have to have *every living person* as a descendant. Could one person really have 7 billion descendants? Yes, indeed. If we assume 25 years per generation, the 2,300-year period represents 92 generations. As long as each of the MRCA's descendants brought into existence, on average, 1.28 persons,[19] the MRCA would end up with more than 7 billion direct descendants.[20]

We don't know whether the MRCA was a man or a woman. It is also possible, by the way, that the MRCA was *both* a man *and* a woman—in other words, that a *couple*, consisting of a man and woman, are the most recent common ancestors of all currently living people. In this case, rather than having *a* MRCA, we would have *co*-MRCAs. To see how this could happen, consider the situation of two full siblings. Ask them to identify their most recent common ancestor, and they will tell you that it is *the couple* consisting of their father and mother.

Before moving on, let me clear up some common misconceptions regarding the MRCA. First of all, she (or he, or they) will not be the common ancestor of every person *ever to have lived*. She will not, in particular, be the ancestor of her own ancestors. She is instead the common ancestor of every *currently living* person. Second, she cannot be the MRCA while she is still alive. Because she is not her own ancestor, she will not, while alive, be the common ancestor of *every* currently living person—including herself. It is only after she has died, and typically only after many generations have passed, that everyone then alive can be her direct descendant. Third, whether you eventually gain the title of MRCA depends on the reproductive activities of other people, something over which you have no control. In particular, if your children don't reproduce, you can never become the MRCA. This means that you gain the title of MRCA not by virtue of being particularly wise or deserving. You instead gain it the same way as you win a lottery—as the result of dumb luck. Fourth, realize that the MRCA isn't the *only* ancestor that every living person has in common. Ancestry, after all, is a transitive relationship, meaning that your ancestors' ancestors are also *your* ancestors. Consequently, the MRCA's many ancestors will also be ancestors of every currently living person. What makes the MRCA special is that, as her name indicates, she is the *most recent* of all these common ancestors.

Although it is unlikely that you will ever gain the title of MRCA, it *is* possible, assuming that you are capable of having children. Suppose, for example, that you and a member of the opposite sex are vacationing all by yourselves on some remote island when the rest of mankind is wiped out by a virus. This event would reset the "MRCA clock." At that point in time, the most recent common ancestor of all currently living people would be the most recent common ancestor of you and your partner, since the two of you would constitute "all currently living people." In particular, if you and your partner are first cousins, the MRCA would turn out to be co-MRCAs—namely, the grandparents that you have in common.

Suppose that while on this island, you and your partner have children. When the two of you pass away, you would become the most recent common ancestors of all currently living people—namely, of all your children. You would thereby gain the title of co-MRCAs and might retain it for very many generations. (A subsequent bottleneck event could take the title from you.) Congratulations would be in order, but of course, you wouldn't be around to receive them.

At this point, a question arises: since it takes two people to make a descendant, would there ever be a *sole* MRCA? Wouldn't there always be *co*-MRCAs instead? In response to this question, consider again our island scenario, but suppose that instead of a couple being stranded on the island, one man and two women were stranded there—let us refer to them as *Al, Barbara,* and *Betty*. Suppose that Al has children by both women and that these children go on to have children. When Al, Barbara, and Betty die, everyone then alive will have Al as an ancestor, but not everyone will have Betty as an ancestor, and not everyone will have Barbara as an ancestor. Consequently, only Al will hold the title of MRCA.

SOMETIMES PEOPLE, ON HEARING THAT we all share a common ancestor, ask the obvious question: if this person (or couple) had died before having any offspring, would our species have gone extinct? It seems like it would. After all, if you remove a person from a family tree, you have to remove his offspring as well: if he hadn't existed, they wouldn't exist. You also have to remove *their* offspring, the offspring of their offspring, and so on, down to the person's present-day descendants. But because we are all descendants of the MRCA, it means that her (or his, or their) non-existence would entail our own non-existence. Consequently, if the MRCA had met with some tragic childhood accident, our species would be extinct, right?

This line of reasoning sounds plausible, but it makes the mistake of assuming that if our ancestors had not found the individuals with whom they in fact procreated, they would not have found someone else. When choosing a mate, you do not sit around and think about all the potential mates who are unavailable because past events prevented them from being born. At least I hope you don't—that would be a recipe for a miserable existence. Instead, you consider the people who *were* born and choose a mate from among them. In particular, if the person who is your maternal grandfather had failed to exist, your maternal grandmother would, in all likelihood, have responded by marrying someone else.

Had the MRCA (or the co-MRCAs) perished in childhood, people would almost certainly still be roaming the earth. They would have someone else as their most recent common ancestor, though. And you wouldn't be roaming the earth with them. Here's why.

Suppose that the person who is in fact your father had died in infancy, and that as a result, your mother married someone else and had a child. This child—let us call him or her *Leslie*—would have different parents than you did, and therefore would have different genes and a different upbringing. It is therefore reasonable to conclude that Leslie would be a different person than you are, meaning that in this alternate-reality scenario, *you* would not exist. More generally, make a change in a person's family tree, and that exact person ceases to exist, although some other person might come into existence. This logic can be extended to show that if the MRCA had perished in childhood, you would not exist—nor would any of the other people who actually do exist. In their place would be people with different family trees and therefore different identities. All these currently existing people would also have someone else as their MRCA.

Of course, in this alternate-world scenario, you would not be around to complain about your non-existence. Notice, too, that you owe your actual existence to a chain of events that prevented many other people from coming into existence: your mother's marrying your father prevented her from marrying the man with whom she could have brought Leslie into existence. I hope you don't feel guilty about this state of affairs. That, after all, would be another recipe for a miserable existence.

3

You Have a Great Throwing Arm

WE HUMANS HAVE ACCOMPLISHED A LOT. We have built skyscrapers and designed computers. We have invented telescopes that let us see distant galaxies, as well as magnetic resonance imaging (MRI) machines that let us see what is inside our skulls. We have composed sonnets, proved theorems, and painted impressionist landscapes. No other species comes close in terms of its accomplishments.

And what is the secret to our success? The obvious answer is that our big brains set us apart. But if brain size was all that mattered, our Neanderthal cousins, whose brains were as big as or even bigger than ours, would be designing cellphone apps alongside us. And sperm whales, whose brains are about six times as big as ours, would have done things that make our own accomplishments look like child's play. On hearing this, we might modify our answer to the what-makes-us-so-great question: it isn't how big our brains are in absolute terms; it's how big they are, relative to our body weight. Yes, sperm whales have huge brains, but they also have huge bodies. We humans, by way of contrast, have pretty big brains in pretty small bodies. The problem with this answer is that in ants, the ratio of brain weight to body weight is far higher than in humans,[1] and again, although they are remarkable little creatures, their list of accomplishments is underwhelming.

It might now be suggested that what makes us special is not the size of our brain as a whole, but the size of our neocortex, the grey matter on the surface of our brain where most higher reasoning takes place. But alas, dolphins' neocortical area is larger than ours, and that of sperm whales is vastly larger. So maybe what makes us special is not the size of our neocortex, but its structure: ours has six layers, compared to a whale's five.[2] Or maybe what makes us special isn't the structure of our neocortex but the structure of the neurons that comprise it.[3] As one reads such suggestions, one can detect what seems like a growing sense of desperation: surely there is *something* unique about our brain to account for the unique accomplishments of our species.

There is reason, though, to think that our success has as much to do with our bodies as with our brains. To see why I say this, consider again sperm whales. Because they have huge brains with impressive neocortices, it is easy to imagine that they are prodigiously intelligent—that as they swim through the ocean depths, they occupy themselves by proving mathematical theorems, thinking up clever gadgets, and composing symphonies. If we encountered one of these "Einstein whales," though, we would be oblivious to his brilliance. Because he is trapped inside a whale's body, he would lack the ability to tell us his thoughts, write them down, or even tap them out on a computer keyboard. He would likewise lack the ability to construct a prototype of the remarkable integrated circuit he has thought up or for that matter, to alter his environment in any significant manner in order to make his life easier. This whale would be like Stephen Hawking sans voice synthesizer. Pity the poor Einstein whale! He has wonderful ideas, but they will never transform the world in which he lives.

I hasten to add that there is no evidence that sperm whales actually have the mental life I just described. I am entertaining this possibility only to demonstrate that although our brain clearly played a key role in allowing us to accomplish the things we have, our bodies also deserve a large part of the credit. Let us therefore turn our attention to the human body and the role it has played in our progress.

ANIMALS CAN DO SOME AMAZING THINGS. They can, for example, see, fly, and echolocate. But there is something even more amazing that we humans can do: walk in our own peculiar fashion. At first this seems like an absurd claim to make, but when we do the math, we find out otherwise.

The ability to see is indeed amazing, but eyes evolved independently on dozens of occasions.[4] The ability to fly is likewise amazing, but it has evolved independently at least four times, in pterosaurs, birds, bats, and insects.[5] (Other animals, of course, can *glide*, including squirrels, "flying" fish, and Chrysopelea, the "flying" snake, but this is different than flying.) The ability to echolocate is rarer, having evolved independently in bats and marine mammals and, in a less sophisticated form, in oilbirds, shrews, and tenrecs. There are nevertheless millions of species that have one of these abilities, and hundreds of thousands that have two out of three of them, including insect species that can both fly and see, and species of marine mammals that can both echolocate and see. And finally, there are more than a thousand bat species that have all three of these abilities: they can see, fly, and echolocate.

Now consider the ability to walk. We humans, to be sure, are not the only animals that can walk and not even the only animals that can walk on two legs. Chimpanzees, bears, and trained dogs can, but only for a short distance; they are also clearly more comfortable walking on all fours. Nor are we the only animals that can run on two legs. Frilled lizards can: their frills lift the front part of their body off the ground when they move fast enough. Likewise, basilisk lizards, also known as *Jesus Christ lizards*, not only can run on two legs but can do so across water. Some cockroach species can also run on two legs.[6] But although these animals can *run* on two legs, they cannot *walk* on two: once they slow down, their front feet return to the earth and they walk on all fours.

We might, at this point, propose kangaroos as an example of an animal that can run on two legs, but it doesn't run, it hops. Many small birds, such as sparrows, also hop instead of run. Although ostriches and roadrunners can run, neither do so with their torso upright, the way we humans do. They instead distribute their weight ahead of and behind their legs, and in doing so resemble a tightrope walker carrying a long pole crossways. This makes it much easier for them to keep their balance.

This brings us to penguins. Thanks to their relatively low center of mass, they can walk—or rather, waddle—in an upright manner that resembles that of humans.[7] When they need to move fast, though, their waddling gives way, if they are on snow or ice, to what is called *tobogganing*: instead of trying to waddle even faster, they lie on their bellies and push with their feet.

What makes human walking ability special, then, is our ability to both walk and run on two legs, to stand upright as we do, and to do so for very long distances, over a wide range of terrains. Not only are we the only animals on Earth with this skill set, but it is likely that the first time it appeared on Earth was in one of our close hominid ancestors. Furthermore, our walking ability is more than just a great party trick in the animal world. It played a very important role in allowing us to accomplish so much. Allow me to explain.

GO BACK 7 MILLION YEARS, and our ancestors would have lived in dense tropical forests. The ability to swing from branch to branch and otherwise travel through the crowns of trees would have been much more important, in survival terms, than the ability to walk. Then a cold, dry spell shrank the forests of the Rift Valley, leaving woodlands that were separated by savanna grasslands, a climate change that might have been caused, in part, by a geological uplifting of that part of Africa.[8] This made it advantageous for our ancestors to supplement arboreal locomotion with an improved ability to walk—and under some circumstances, run—from one wooded area to another. The timeline for this change is unclear, but we do know that by 3.5 million years ago, our ancestors had the ability to walk upright rather than knuckle walking, the way modern chimpanzees do. The most striking evidence for this transition is the remarkable human footprints found at Laetoli in northeastern Africa. It is also significant that the individuals who made these footprints had big toes that were more like the in-line toes of modern humans than the out-turned big toes of modern chimpanzees (see Figure 3.1).

The ability to walk upright would have benefitted our ancestors in a number of ways. When moving from woodland to woodland, walking upright, even for short distances,

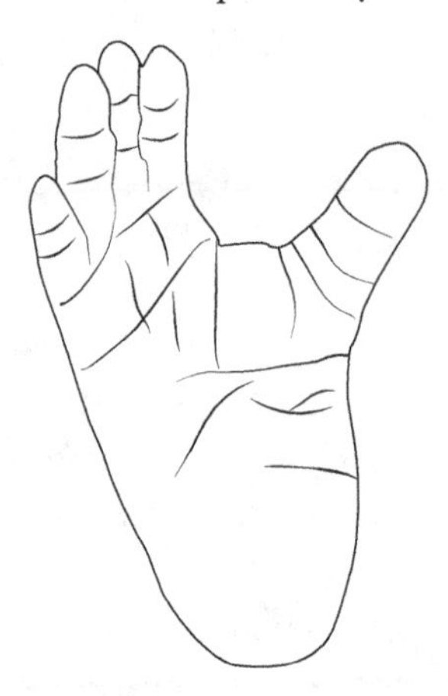

FIGURE 3.1. The foot of a chimpanzee. Notice the out-turned big toe.

would have increased their chances of spotting predators in the grass. Bipedal walking is also more energy-efficient than walking on four legs the way cows do or knuckle walking the way chimpanzees do.[9] And when, over the course of generations, they became proficient at walking upright, our ancestors could have used their hands and arms to carry things, including food for their family and tools they had made. Those with the greatest ability to walk upright would have been most likely to survive and reproduce, and their offspring would have inherited their parents' bipedal ability.

Our arms, by the way, aren't useful just for carrying things. They also play an important role in letting us run in an upright position. Impair our ability to swing our arms as counterweights, and our running ability is impaired. This is part of the reason why police, when they arrest someone, cuff his hands behind his back: besides making it harder for the person to strike them, it makes it harder for him to outrun them.

Becoming bipedal changed our ancestors' bone structure. It also resulted in their gaining muscle mass in their legs and losing it in their arms. A chimpanzee that spends hours a day climbing through trees might have 36 percent of his muscle mass in his arms. We humans have only 20 percent, but our legs are more muscular than those of chimpanzees.[10]

In a tropical forest, food would have been readily available to our ancestors, meaning that there was little need for their bodies to store fat. When our ancestors ventured forth from the forest, though, their food supply became less reliable. There would have been days in which nothing edible could be found, and there would have been seasons that had many such days. Our ancestors adapted to these circumstances by storing fat. That way, they would have calories to burn when food was scarce. This adaptation is reflected in the body composition of modern humans. A female who is slender (by current standards) might be 30 percent fat by body weight; a female chimpanzee, by way of contrast, might be only 3.6 percent. And whereas a slender (by current standards) human male might be 20 percent fat by body weight, a male chimpanzee might be only 0.005 percent! Even a human who died from starvation would have more fat than that.[11]

Realize that reserves of fat aren't just unnecessary for chimpanzees; they are a burden. It is much easier to move weight around horizontally than to move it vertically, against the force of gravity. Consequently, although a person who is a hundred pounds overweight will still be able to walk around and even run, he will be hard pressed to climb a tree and, once in that tree, to pull himself to higher branches, the way chimpanzees do. Because our ancestors abandoned trees for the ground, they could afford to carry extra fat.

THE ABILITY TO SPOT PREDATORS and to carry things weren't the only benefits of walking upright. When they stood up, our ancestors' bodies were farther from the sunbaked ground. They were also less exposed to the sun. (Sun bathers, of course, lie down to increase this exposure.) Furthermore, their upright torsos had more "sail area" than if they walked on all fours and were therefore better cooled by any breezes that might be present—and if nothing else, by the slight breeze they created for themselves by walking.

Their ability to stay cool under the midday sun was enhanced by other changes. About a million years ago,[12] they not only lost their heat-trapping fur, probably in stages, but gained the ability to sweat in response to heat. (Chimpanzees have sweat glands under their fur but do not sweat when they get hot.[13]) This ability came at a price—humans had to drink lots of water to replace the sweat they lost—but because walking upright freed their arms to carry things, they could have used ostrich eggs or hollow gourds as canteens.

Losing their fur made our ancestors, as we have seen, susceptible to skin cancer, but it also heightened their sense of touch. One imagines that touch would therefore have become an important form of nonverbal communication. For modern humans, it can be a delicious and even electrifying experience to have someone touch our skin. It is not unreasonable to think that our relatively hairless ancestors would also have enjoyed this experience.

Our ancestors, then, were probably the coolest characters on the savanna. While other animals would quite sensibly spend the afternoon taking a nap in the shade of a rock or tree, our furless, sweaty, upright ancestors might have been walking or even running about. In this respect, they would have resembled the modern Afar tribesmen of Ethiopia who will, without thinking twice, walk 25 miles (40 kilometers) in blazing heat, chat with someone for half an hour, and then walk back home.[14] This tolerance for the midday heat would have given them an edge in their pursuit of game. Although their prey might have been faster than they were, it would have been a furry, sweatless quadruped and therefore would have lacked their stamina, meaning that our ancestors could run it to death.

This sort of persistence hunting is still practiced. The Kalahari tribesmen of southern Africa follow an animal—an eland or kudu, perhaps—for several hours until it literally cannot take another step. They then dispatch the helpless creature. Paleoanthropologist

Louis Leaky wasn't an African tribesman, but he nevertheless inherited the endurance of the ancestors he had in common with them. On one occasion, in order to win a bet, he ran down an antelope. The antelope would sprint away whenever Leakey got close, but then it had to stop and pant in order to cool down. Leakey just kept coming at a slow, steady pace, his body cooled by sweat. In the end, the animal succumbed to heat stroke.[15]

GO BACK FAR ENOUGH in your ancestry, and you will come to creatures who used their arms to move up, down, and through trees. Although we humans rarely use them for this purpose, we have taken advantage of their remarkable range of motion. They can point in any direction away from our body, except the space occupied by our head and torso—although admittedly, the space directly behind our back can also be tricky. They can also swing in complete 360-degree circles in many different planes. Our hypermobile arms differ from those of our great ape cousins in one important respect, though. Stretch your arm straight out in front of you with your palm facing down, and then start rotating the arm up and over your shoulder, until it is as far back as it will go. As you do this, you will feel the muscles and tendons in that shoulder tightening. Energy is thereby stored, and when the arm is subsequently moved forward, that energy is released.

Think about the motion of a baseball pitcher. As he winds up for a pitch, he stores energy in his upper arm. He also rotates his torso, something other great apes have trouble doing.[16] As he moves his hand forward, he "unwinds" his torso and steps into the throw. As a consequence, a baseball might leave his hand at 100 mph. To put this feat into perspective, a chimpanzee might be able to manage only a 20-mph throw.[17]

And besides being remarkable for their speed, our throws are remarkable for their accuracy. A chimpanzee is lucky to have a thrown object go in approximately the intended direction. (The exception, I am told, is if he is flinging feces at some unlucky researcher.) A big-league pitcher, by way of contrast, can consistently throw strikes. This accuracy is made possible by two things. The first is having hands with fingers that are shorter than those of chimpanzees (see Figure 3.2).[18] The second is having the ability to time the release of a ball to within 0.5 milliseconds.[19] A pitcher who lacked this timing would throw balls instead of strikes. The combination of speed and accuracy means that we humans have what is by a large margin the best throwing arm in the animal world. Even if you are old and arthritic, there is a good chance that you can outthrow a chimpanzee.

For a long time, our ancestors got their meat primarily by scavenging and by hunting small animals. About 2 million years ago, though, they started hunting larger game, an undertaking that would have been both difficult and dangerous. If they directly assaulted an animal by running up and bashing it with a rock or repeatedly clubbing it, they ran the risk of getting kicked, bitten, gored, or trampled. A major breakthrough came when they figured out how to use their throwing ability to kill animals at a distance. They presumably first used this ability to hit animals by throwing rocks[20] or sticks, then by throwing sharpened rocks or sticks, and finally by throwing sharpened rocks tied to sticks. This

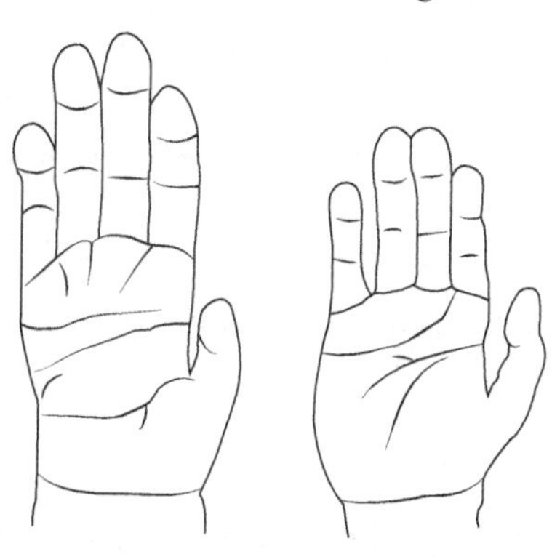

FIGURE 3.2. A chimpanzee hand and a human hand, drawn to scale.

allowed them to take down animals much bigger than themselves; indeed, when our human ancestors appeared in some corner of the globe, it typically wasn't long before the local megafauna went extinct.

After they had killed an animal, our ancestors' throwing ability would have made it easier for them to retain ownership of its carcass. In battles over a carcass, it wouldn't have been necessary for them to kill scavengers; all they had to do was cause them pain. This could be accomplished with an accurately thrown rock. Likewise, the ability to cause pain at a distance might have allowed our ancestors to steal carcasses from other predators. And finally, despite their shorter fingers, our ancestors would have retained much of their tree-climbing ability, meaning that they might not even have had to throw rocks to steal a carcass. They could instead have simply waited until a leopard, after stashing its kill in a tree, went off to attend to other business. When it returned, its planned lunch, or large parts of it, would be mysteriously missing.[21]

Evidence suggests that archaic humans started exploiting their ability to kill at a distance about 2 million years ago.[22] This ability was enhanced by the invention, in the last 100,000 years, of the atlatl and the bow and arrow.[23] And in this century, we have developed drones that allow someone, sitting in a comfortable environment, to track down and kill people who are thousands of miles away.[24]

Fingers are obviously necessary for the controlled throwing of a projectile. I had assumed that thumbs are as well, but it turned out that I could throw fairly far and accurately without using my thumb. And presumably, with practice—in particular, if I had engaged in thumbless throwing since childhood—I would be as accurate throwing things without using my thumb as I am throwing things using it.[25]

Although you can throw things pretty well without opposable thumbs, the same cannot be said of clubbing things. An opposable thumb can be wrapped around the handle of a club, thereby vastly reducing the chance that you will lose hold of it at impact, meaning that you can continue to club your target.[26] Possession of an opposable thumb also enables us to make a fist—we are the only animal that can do so—which in turn enables us to punch things.[27] And not to be forgotten, the possession of opposable thumbs allows us to get a precision grip on objects. In particular, it is indispensable for doing the manipulations necessary to make and use flint knives or blades, and to fashion spears. There are lots of things that you can do easily if you have thumbs that you will do clumsily, if at all, without them. One way to gain an instant appreciation of your thumbs is to tape the top segment of them to the bottom segment of your index fingers. I tried this experiment. The result: instant klutz.

BESIDES HAVING A GREAT THROWING ARM, you possess a remarkable vocal apparatus. At its heart lie vocal cords, but significantly, you also have a jaw, tongue, and lips that can be moved in many directions to shape the sounds those cords make. And not to be forgotten, you have the ability to expel air from your lungs for an extended period and with precise control. Indeed, when you are about to utter a long sentence, you unconsciously prepare for it by inhaling enough air to complete it. Because you have these abilities, you can, when circumstances call for it, communicate with a barely audible whisper or a theater-filling song or proclamation. In terms of your vocal ability, you are a freak of nature.

To be sure, we are not alone in our ability to make sounds. Crickets and snapping shrimp make noises with their appendages. Madagascar hissing cockroaches make them by expelling air from their abdomen. Snakes make a similar hissing noise with their mouth and in some cases can also make a rattling noise with their tail. And of course, animals with vocal cords can vocalize: lions roar, humpback whales sing their beautiful songs, and we humans chatter.

Some animals are capable of making human noises. It is possible, on hearing a parrot utter "Hey, you," to think that some person is trying to get our attention. Because they lack our breath control, though, parrots can't make the sustained sounds that we can, and as a result, if a parrot tried to recite, say, the opening sentence of Lincoln's Gettysburg Address—"Four score and seven years ago, our fathers brought forth on this continent . . ."—no one would be fooled. Significantly, chimpanzees have a similar shortcoming. Although their vocal apparatus resembles ours, they lack our breath control, and this has a dramatic impact on the sounds they can make.

We humans presumably gained our linguistic ability in stages. Our first "language" would have been "body language." When angry or afraid, for example, the fur of our hominid ancestors would involuntarily have stood up to make them look bigger and thereby intimidate whatever it was that angered or scared them. We still have this body language wired into us, but because we have lost our fur, the language in question is muted: scare or anger us, and we might respond by getting goose bumps—which, of course, have zero power to intimidate.

Our ancestors' bodies would also have "spoken," in a semi-voluntary manner, with their posture and gait. And because they had arms, they could make gestures to indicate their mental state. A gesture might have involved waving their arms in the air or pointing at some object, something that a rhinoceros can't do. Furthermore, because of their facial anatomy, they could communicate with facial expressions, something, once again, that a rhinoceros can't do.

Noises probably accompanied our ancestors' gestures, as a way to make others look at them while they were gesturing. If a certain noise routinely accompanied a certain gesture, the noise itself could come to stand for that gesture. This would have come in handy when our ancestors had to communicate in dense foliage or in the dark. Spoken language of a primitive sort could thereby have arisen, and with the passage of time, the range of meaningful noises would have expanded and the noises in question would have become more specific and would have conformed to certain rules.

Our vocal apparatus, by making complex language possible, allowed us to coordinate our increasingly complex plans with those of other people. This in turn made it possible for us to accomplish things in groups that we would have been unable to accomplish by ourselves. In particular, it would have enhanced our hunting ability. It is one thing for a game animal to evade a single person with a spear; it is quite another for it to have to deal with spears thrown from all directions, as might happen in an ambush. Such an ambush, though, would be difficult to plan without language.[28]

More generally, language makes large-scale cooperation possible. By way of illustration, imagine a group of engineers trying to design and construct a jumbo jet without using language. For that matter, imagine two couples trying to plan, without using language, something as simple as a joint dinner, at 6 pm next Saturday, in which one couple provides the salad and dessert, and the other provides the entrée. It would be an impossible undertaking.

BESIDES REQUIRING THE USE OF LANGUAGE, cooperation requires self-control. It also requires a trusting nature and, in particular, a willingness to trust complete strangers, to a degree. Along these lines, consider what happens when you fly somewhere. You and hundreds of other people you have never met board a plane in an orderly manner. You all find your seat and quietly occupy it for the duration of the flight, at which point you all exit the plane, again without incident. Try transporting a few hundred chimpanzees in this manner, and mayhem will result. Likewise, when we are walking through most parts of most cities, we ignore the strangers we encounter because we assume that they will do us no harm. Forest-dwelling chimpanzees that were similarly trusting would likely pay a price.

Another indication of the cooperative nature of our *Homo sapiens* ancestors is their possession of objects, including seashells and flints, that could only have come from a distant source. It is possible that a single individual transported these goods, but it is much more likely that it was the result of a trading network. Consider seashells, for example. A person who lived at the seashore would have them in abundance. He could have traded them with someone who lived a bit inland and thereby gained access to goods not available at the seashore. That trading partner could in turn have traded them with someone who lived even further inland, and so on. In this manner, the seashells could have traveled hundreds of miles inland without any one person having carried them there. Notice, too, that a trading network of this sort is possible without anyone setting it up and without anyone subsequently maintaining it. To the contrary, it arises and functions in a spontaneous manner. The shells are drawn inland, as it were, by human desire.

Neanderthals do not appear to have been as actively involved in trade as their *Homo sapiens* cousins were. This is unfortunate, since their reluctance to trade deprived them not only of the goods available through trade but of the information that could be obtained from their trading partners. It thereby made it harder for Neanderthals to compete with our species.[29] And why were the Neanderthals reluctant to trade? Quite possibly, it was because they lacked the trusting nature of their *Homo sapiens* cousins.

TO SUMMARIZE, our hominid ancestors had complex brains, housed in a rather unusual body, and as a result, they gained the ability to kill at a distance. This was, to be sure, an accomplishment, but it pales in comparison to what we have subsequently gone on to accomplish. Something must have happened to put us on the fast track to greatness.

Anthropologist Richard Wrangham has argued that as a result of our improved killing ability, our ancestors' consumption of meat increased dramatically, and their bodies responded to this improved diet by growing brains that were even more complex. With their enhanced brainpower, they could think up new ways to get even more meat, more reliably. They thereby found themselves revolving around a virtuous circle that benefitted them enormously.

By including meat in their diet, our ancestors were able to get their calories much more efficiently. By way of illustration, you can get 213 calories from 85 grams (about 3 ounces) of 85 percent lean ground beef; to get this many calories from spinach, you would have to eat 900 grams (about 2 pounds). That same 85 grams of beef will provide you with 22 grams of protein; to get this much protein from spinach, you would have to eat 750 grams (about 1.6 pounds), and it would be lower quality protein. To meet their nutritional needs, then, our primarily vegetarian ancestors would have had to eat a lot of fruits and vegetables, and this would have been a time-consuming process. Whereas modern humans might spend an hour a day eating, our ancestors might, like chimpanzees, have spent six.[30]

In nutritional terms, brains are expensive organs. Growing them requires lots of fat and protein, and operating them burns lots of calories: even when you are at rest and not thinking particularly hard, your brain, which represents 2 percent of your body mass, might consume 20 percent of the calories you burn. When our ancestors started eating more meat, though, they could "afford" bigger, more complex brains. Furthermore, about 2 million years ago,[31] our ancestors became smart enough to start and maintain the fires necessary to cook meat. This in turn increased their ability to extract the nutritional value from that meat.[32] As a result, the virtuous circle described above accelerated, and the pace of human progress shifted into high gear.

So why has our species been able to accomplish so much? Yes, our brain played a key role, but so did our body. It was furless and could sweat. It could walk and run in an upright position. It had a certain kind of arms, hands, fingers, and thumbs; a certain vocal apparatus; and lungs with superb breath control. And it is significant that our brain, besides being smart, was wired so we would have a cooperative personality. Remove any of these characteristics, and it is entirely possible that we would not today be writing or reading books about our extended ancestry; we would instead be out foraging, scavenging, or hunting for our next meal.

It is, to be sure, a long and curious path that leads from the possession of a great throwing arm or an ability to walk upright to the invention of the Large Hadron Collider, but it is the path our species took.

4

Your Place on the Tree of Life

CONSIDER THE *TYRANNOSAURUS REX,* affectionately known as Sue, whose bones are proudly displayed in Chicago's Field Museum. Sue was named after Sue Hendrickson, the paleontologist who discovered her—or maybe him, since the dinosaur's sex isn't clear. In her prime, Sue (the dinosaur) must have been a magnificent beast. Realize, though, that you and she have an ancestor in common, meaning that you are cousins, in an extended sense of the word. This common ancestor lived 320 million years ago. You also have ancestors in common with any other organism, living or dead, that you care to name, including a penguin, a sequoia, and the *E. coli* bacteria that roam your intestines. And the organisms to which you are related are in turn related to each other. Once we acknowledge that all species are related, we can construct a tree of life that displays these relationships.

We have already explored the logic of family trees. They have persons as their "nodes." Each node has a *T* above it, with the person's parents at the ends of the crossbar of that *T*. These parents in turn have *T*s above them. Every part of a family tree will have this same shape, and as a result, it will look like a diagram an integrated-circuit designer might draw (see Figure 4.1). Furthermore, the family trees of any two people will look exactly alike, with *T*s stacked on *T*s. The only difference will be the identities of the individuals who appear on their trees—unless, of course, the two people are full siblings, in which case their family trees will be identical.

The tree of life, by way of contrast, is concerned not with people but with species. The complete tree of life would resemble an elm. At the bottom of the tree would be the "roots." They represent the lifelike but not fully living things that gave rise

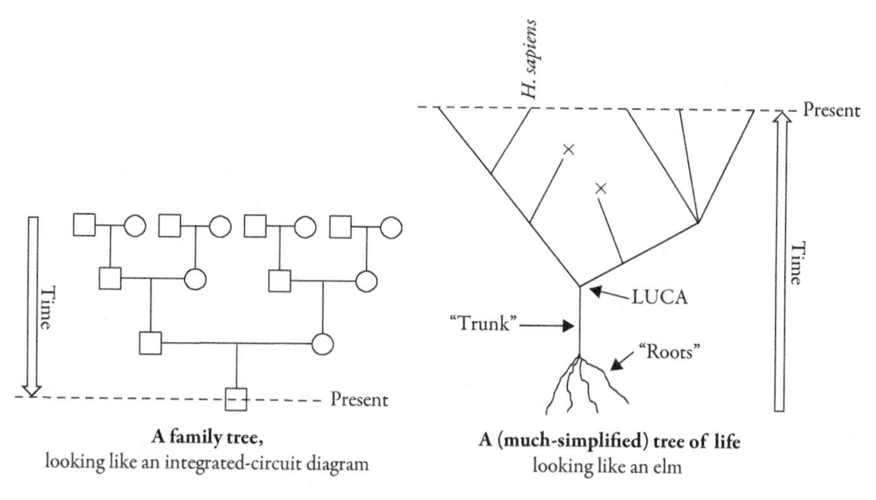

FIGURE 4.1. Whereas family trees show how people are related, the tree of life shows how species are related. Notice that the "arrows of time" for these trees point in opposite directions. Notice, too, that in the (vastly simplified) tree of life, many of the branches extend to the present time, resulting in a "flat top." These represent species that still exist. The branches marked with *X*s represent species that went extinct. LUCA is the Last Universal Common Ancestor, the most recent direct ancestor of all currently living things. The roots of the tree represent the semi-animate things that gave rise to the first living organisms.

to the first living organisms. Above the roots would be the tree's "trunk," and at the top of that trunk would be LUCA, the Last Universal Common Ancestor of all currently living things. (We must be careful not to confuse LUCA with MRCA, the Most Recent Common Ancestor of all currently living people, that we discussed in chapter 2.) The tree of life has a trunk because it would have taken time for evolutionary processes to transform the first very simple living organisms into the rather more complex LUCA. At the very top of the tree will be a "twig" for every currently existing species, including our own. Beneath these twigs will be "branches" that lead back to LUCA, which is the common ancestor of them all. See Figure 4.1 for a much-simplified version of this tree.

The "arrow of time" on a family tree customarily points down, meaning that as we move up the tree, we move back in time, to ever-more-distant ancestors. In the tree of life, though, the arrow of time customarily points up, meaning that currently existing species will be at the top. Because there are so many extant species, the top of the tree of life will be broad and flat, more like that of an acacia tree than an elm. And besides these branches, there will be very many that don't make it to the top. They represent species that went extinct. Sometimes many species go extinct at about the same time in what is known as a *mass extinction event*. These events will appear on the tree of life as flat-topped "terraces" within the tree, as can be seen in Figure 4.2.

We can make our tree of life more informative by having the horizontal distance between branches correspond to how far apart species are, in genetic terms. In such a tree,

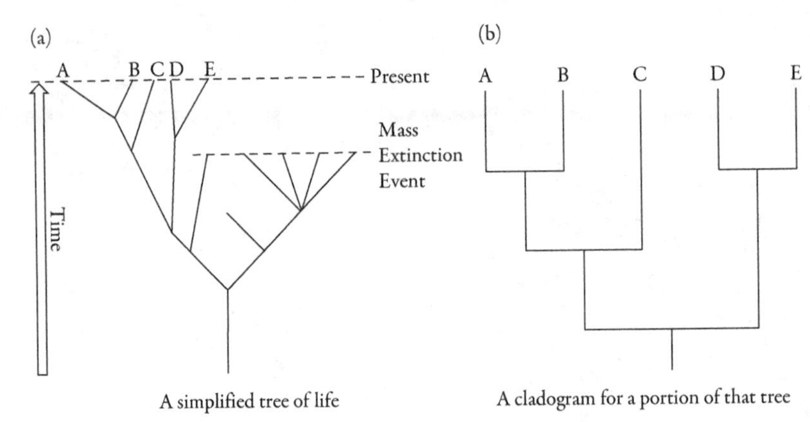

A simplified tree of life

A cladogram for a portion of that tree

FIGURE 4.2. On the left is a tree of life showing an extinction event. The flat-topped "terrace" within that tree indicates the nearly simultaneous extinction of many species. On the right, a portion of tree (a) has been converted into a cladogram for species A through E. Cladograms are often shown tipped sideways to make labeling easier.

a species that stayed the same over a long period would be depicted by a vertical branch. This is how the branch depicting the coelacanth, a "living fossil" that has remained much the same for 400 million years, would look. Species that changed with the passage of time would be depicted by slanting branches.

We can also make our tree of life less informative but significantly easier to construct by concerning ourselves not with how long ago species split apart or how far they have drifted, genetically speaking, but simply with which species split from which and the order in which they split. The resulting diagram will be what biologists refer to as a *cladogram* (see Figure 4.2). And finally, we can save space on a page by constructing radial trees of life (see Figures 4.3 and 4.4). Despite how densely packed these radial trees are, they display only a tiny portion of the species that exist. It is instructive to compare our entirely unprivileged place on these modern radial trees with our place atop the rather ornate tree of life constructed by nineteenth-century biologist Ernst Haeckel (see Figure 4.5).

We have seen that a complete family tree would be very wide. The same will be true of a complete tree of life. There are perhaps 10 million currently existing species, most of which have not yet been discovered.[1] Suppose, then, that we made a complete tree of life and allowed a centimeter between the top twiglets of this tree. The piece of paper we constructed it on would have to be 100 kilometers—more than 60 miles—wide. But realize that a truly complete tree of life would show not only all the species that currently exist but also all the species that have ever existed. It is likely that 99 percent—or maybe even 99.9 percent—of species ever to exist have gone extinct,[2] and all these species would have to be included in the tree of life as branches that didn't reach the flat treetop. The resulting tree would be densely branched and enormously complex.

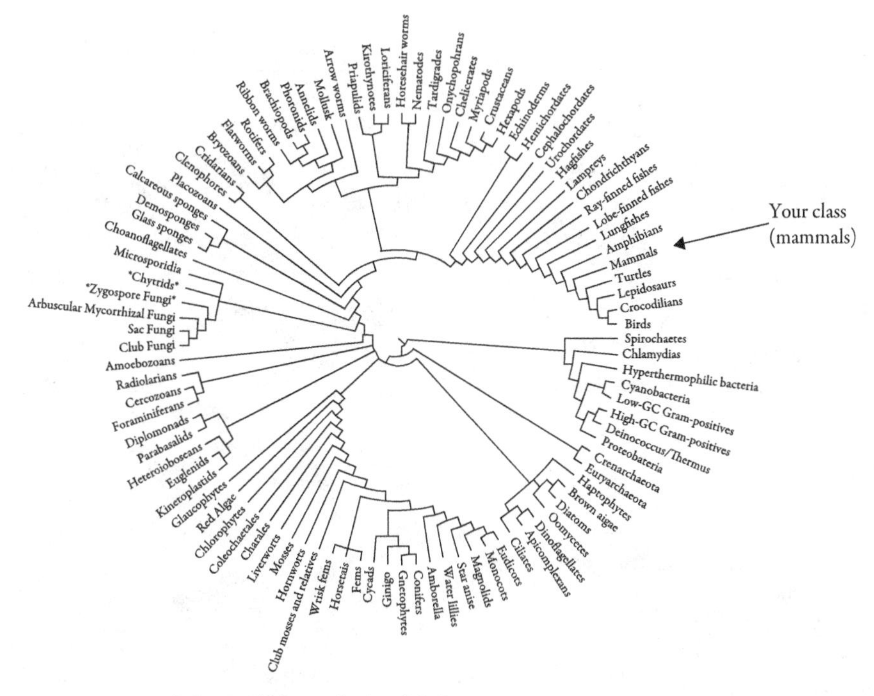

FIGURE 4.3. A radial tree of life, vastly simplified.

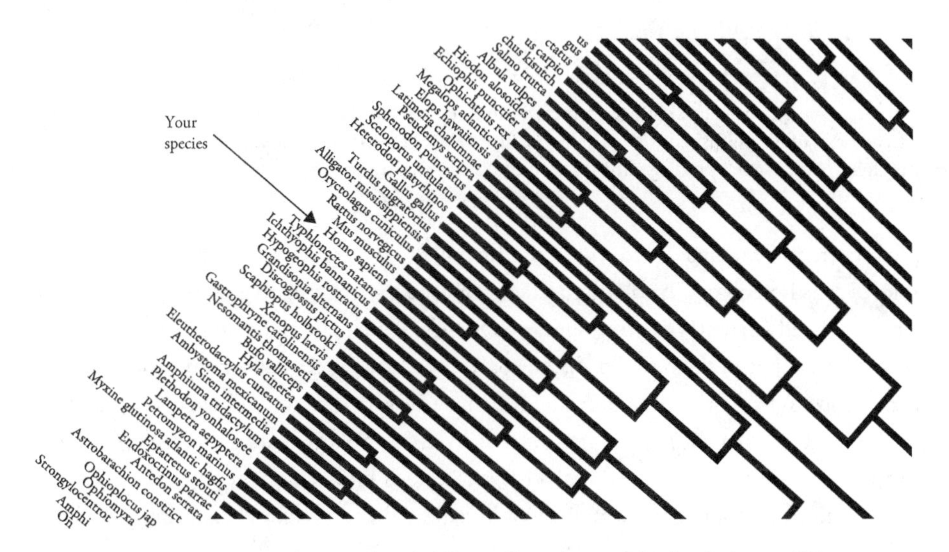

FIGURE 4.4. One portion of a more detailed (but still vastly simplified) radial tree of life.

TO CONSTRUCT A FAMILY TREE, we need to know who lived in the past, along with who among them begat whom. Similarly, to construct the tree of life, we need to know what species have existed and how these species are related, in evolutionary terms. It turns out, though, that we are sadly lacking in both sorts of knowledge. We have yet to discover

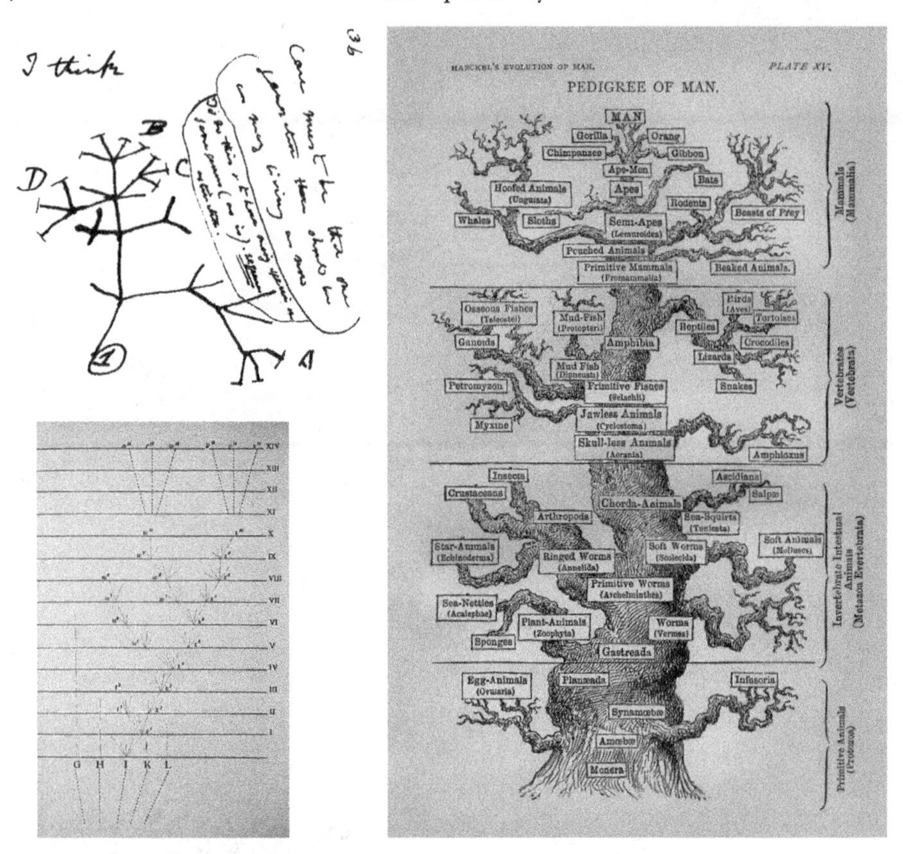

FIGURE 4.5. Historically significant trees of life. Charles Darwin's "aha-moment" tree is at the top left, and one of his published trees is below it. Ernst Haeckel's rather more elaborate tree of life is on the right. Notice man's privileged position at the top of this tree.

most currently existing species, and the further back in time we go, the sketchier our species census becomes. And although we can use DNA testing to gain insight into how closely related currently existing species are, and how closely they are related to species that went extinct in the not-too-distant past, the older the remains of an extinct species are, the less reliable the tests become. It seems unlikely that we will ever be able to do testing on the remains of an organism that lived 10 million years ago—the DNA will simply be too far gone to decipher. This is a serious limitation when we consider that construction of a complete tree of life would require that we tell how organisms that lived billions of years ago were related.

The problems that arise in constructing the tree of life become apparent as soon as we investigate our own portion of that tree. We might think that, given human narcissism, we would have a deep understanding of how the hominid species are related, but we find just the opposite. Scientists argue about what hominid species have existed; when they agree that a species existed, they argue about when it existed; and finally, they argue about how the species that existed are related. Paleoanthropologist Ian Tattersall, who has been a participant in this debate throughout his career, says that in the last two decades, he has gone from thinking that there were a dozen hominid species that evolved over a period of 4 million years to thinking that there were two dozen species that evolved over a period of 7 million years.[3]

The ideal way to construct the tree of life would be to travel back in time. We could start our research by tracing the evolutionary origin of our own species, and one way to do this would be to construct an extended version of our family tree. As we traveled back, we would keep track not only of who the parents of our ancestors were, who the parents of those parents were, and so on, but at each stop, we would make a note about what species an ancestor belonged to. If we traced our ancestry back 500,000 years, our direct ancestors would be members not of our own species *Homo sapiens* but members of the species *Homo heidelbergensis*—if, at any rate, current consensus paleoanthropological views are correct. We would conclude that *Homo heidelbergensis* is our "parent species," and in the tree we constructed, we would show our species branching from its branch. If we traced our ancestry back another million years, we would find—again, if consensus views are correct—that members of the species *Homo erectus* are the direct ancestors of members of the species *Homo heidelbergensis*, making *Homo erectus* our "grandparent species."

Continue back 7 million years and we arrive at the species that was the common ancestor of humans and chimpanzees. Members of this ancestral species would have been neither humans nor chimpanzees. They would instead belong to a species that no longer exists. And going back 30 million years, we would find direct ancestors that belong to the now-extinct ancestral species that we have in common with monkeys. Bottom line: you definitely aren't a monkey's uncle, but you *are* a monkey's cousin, in some very extended sense of the word.

Let me pause here to respond to what creationists sometimes take to be a knockout-punch criticism of evolution: "If man evolved from monkeys, how come monkeys are still around?" The answer to this question has two parts. First, we didn't evolve "from monkeys"; we and monkeys evolved from a common ancestor that was neither a human nor a monkey. Second, even if our species *had* branched off from monkeys, it is entirely possible for monkeys to have survived the branching. By analogy, although Protestants branched off from Catholics, there are still lots of Catholics around.

ALTHOUGH IT IS HIGHLY UNLIKELY that we will ever possess a complete tree of life, we do have a general understanding of how our species came to exist. As I've said, *Homo heidelbergensis* and *Homo erectus* are probably our parent and grandparent species, respectively. Our "great-grandparent" species, which belonged to the genus *Australopithecus*, appeared 4 million years ago, and a million years before that, we find a species belonging to the genus *Ardipithecus*. Go back even further on the tree of life and we come, as I have said, to the ancestor we share with chimpanzees and then to the ancestor we share with monkeys.

Press on and we come to early mammals, many of which resembled modern lemurs. Go back 66 million years, and we arrive at the event that played a pivotal role in our evolution: an asteroid struck the earth and took out the dinosaurs.[4] Mammals fared somewhat better, perhaps because they had underground burrows, the way modern moles do, or spent long periods hibernating underground, the way modern tenrecs do.[5] This discussion makes it clear that an event that has catastrophic consequences for one group of organisms can be a blessing for another. Thanks to an asteroid having struck the earth, our mammalian ancestors found themselves living in a world with fewer competitors for the available food and likely fewer predators as well. Had the asteroid not struck, we might not exist.

The End-Cretaceous extinction event just described is the most famous of the five mass extinction events that biologists have identified, but it was not the biggest. Whereas the End-Cretaceous took out an estimated 75 percent of species, the End-Permian event 252 million years ago is estimated to have taken out 96 percent. It is thought to have been the result of a combination of events, including an asteroid strike and large-scale volcanism.[6] The three other mass extinction events were the End-Triassic, 200 million years ago; the Late-Devonian, 375 million years ago; and the End-Ordovician, 445 million years ago.

Reading this list gives rise to two questions. First, why do these events take place at the ends of geological eras, such as the Cretaceous or the Permian? It is because geologists divide time into eras largely on the basis of what fossils they find, which in turn is profoundly affected by mass extinction events. Second, why are there no mass extinction events before 445 million years ago? It isn't because the earth went 4 billion years without asteroid strikes, volcanic events, or extreme climate changes. Many such events doubtless took place. Indeed, according to the Snowball Earth hypothesis, about 650 million years ago, the earth experienced an ice age that left it almost completely covered with ice.[7] Such an event would have had a major impact on whatever was then alive, but because the life forms that then existed were not susceptible to fossilization, we have no record of a mass extinction event, meaning that it doesn't make our list.

GO BACK EVEN FURTHER on the tree of life and you find, in order of increasing antiquity, mammal-like reptiles, reptiles, and amphibians. Go back a bit further and you come to a key event in your extended personal history. There must have been a point at which your fish ancestors made the transition to land; otherwise, you wouldn't be a land animal. But how could something that lives in the water transition to land?

When you think of a fish making this transition, it is tempting to imagine it crawling out of the ocean onto a sandy or pebbly beach. A case can be made, though, that the first fish to leave the water would not have crawled; it would instead have flipped onto a rocky shore, the way modern leaping blennies do, perhaps in order to evade a predator. Fish that lived in the shallow, poorly oxygenated water of swamps and marshes might also have made the transition to land. A fish species that was accustomed to crawling on fins underwater to graze might, over many generations, have developed the ability to crawl briefly onto land, perhaps in its pursuit of food or perhaps to lay its eggs in a place that would be safe from predation.

This brings us to Tiktaalik, the fossil that made headlines after its discovery in 2004. News coverage of the event created the impression that it was the first fish to leave the water. And since Tiktaalik was found on a hillside on Ellesmere Island in the Arctic Ocean, its discovery also gave rise to the rather incongruous mental image of a fish leaving the ocean to climb a nearby hill, only to die, presumably as the result of hypothermia, and thereby become a fossil for us to find.

This image is dispelled when we realize that 375 million years ago, when Tiktaalik lived, the place where it was found was a tropical flood plain. Furthermore, it isn't necessarily the case that Tiktaalik was on dry land at the time of its death. It is much more likely that it died in the water and was entombed in mud that subsequently hardened and was uplifted by geological forces. Nor was Tiktaalik a fish; it was instead somewhere between fish and tetrapods on the tree of life.

When the Tiktaalik discovery was announced, some people, besides thinking it was the first "fish" to leave the sea, also got the idea that Tiktaalik was their direct ancestor, in the same way that their mother's father's father is. They thought, in other words, that if they constructed their hyperextended family tree, *that very creature* would appear on it. It is highly unlikely, though, that this is the case, just as it is highly unlikely that a randomly chosen dead person—say, Edward Barton, the first prime minister of Australia—is on your ancestral family tree. Indeed, for all we know, the organism that became the Tiktaalik fossil had no offspring, meaning that it wouldn't appear on *any* currently existing organism's family tree.

But there is another sense in which Tiktaalik could be our ancestor. Even though the organism that became the Tiktaalik fossil isn't a direct ancestor of ours, that organism's species *Tiktaalik rosea* might be a direct ancestor of our species *Homo sapiens*, the way *Homo heidelbergensis* and *Homo erectus* are, but much further back on the tree of life than they are. This is indeed possible, but again, there is as yet no evidence that it is the case. It is much more likely that *T. rosea* is merely a distant relative of our species, the way *T. rex* is.

At this point, it is important to realize that your family tree and the tree of life are connected. In particular, if *T. rosea* were indeed an ancestral species of *Homo sapiens*, at least one member of the species *T. rosea* would have to appear on your family tree. Conversely, if at least one member of the species *T. rosea* appears on your family tree, then *T. rosea* would have to be a direct ancestor of your own species, *Homo sapiens*, on the tree of life.

More generally, keep in mind that as you explore the tree of life, there will, at any point in time, *be exactly one species that is your direct ancestral species.*[8] By way of illustration, consider again the asteroid strike that killed the dinosaurs 66 million years ago. As we have seen, that strike left many mammalian species alive. Among them was the one and only species that, at the moment the asteroid struck, was your direct ancestral species. Furthermore, some members of that species—but probably not all of them[9]—will necessarily appear on your hyperextended family tree; otherwise, that species wouldn't be an ancestral species of yours.

Moving further back into your extended ancestry, we find eels, and before that, worms. Before 535 million years ago—the time of the so-called *Cambrian Explosion*—your ancestral trail gets harder to follow. This is because your ancestors who lived then would have lacked bones and shells, making it difficult for them to leave fossils. There are some exceptions, though, such as the Ediacara biota, discovered in the Ediacara Hills of Australia's Flinders Ranges. These organisms lived during the hundred million years before the Cambrian Explosion. The soft-bodied Francevillian biota, discovered near Franceville, Gabon, are another exception (see Figure 4.6). These organisms, which lived 2.1 billion years ago,[10] look quite unlike any modern organism, and yet, it is possible—but

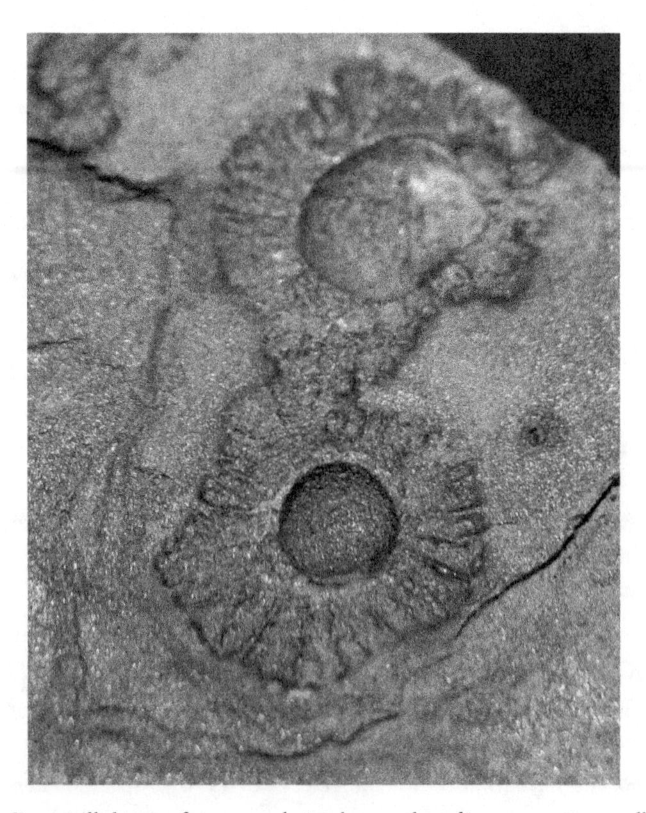

FIGURE 4.6. Franceville biota—for sure a relative, but maybe a direct ancestor as well?

not likely—that some of them are your direct ancestors, in which case they would appear on your very extended family tree.

Going back further in time, all living things were microbes. Your direct ancestors would have been among them. It is, to be sure, difficult for microbes to leave direct fossil evidence of their existence, but they can leave indirect evidence. By way of illustration, consider stromatolites, the rocks "built" by cyanobacteria that did not themselves leave fossils. Some of these rocks date back 3.5 billion years. (Stromatolites, by the way, are still being formed, most famously in Shark Bay in Western Australia.) And besides creating stromatolites, ancient cyanobacteria produced abundant oxygen which affected the chemical makeup of the rocks in the earth's crust. As a result, even though an ancient rock doesn't contain any fossil evidence of life, the chemistry of that rock can provide compelling evidence that life existed when the rock was formed.

WE HAVE SEEN THAT EXPLORING your family tree can turn up surprises. So can exploring the tree of life. It is a bit unsettling to realize that some of our ancestors were squirrel-like creatures, and even more so to realize that others were literally worms. It is likewise surprising to discover that hummingbirds and crocodiles are related, as are butterflies and lobsters. And that the chicken we might have eaten for dinner is a direct descendant of dinosaurs is thought-provoking: oh, how the mighty have fallen!

Before ending this chapter, let us take a moment to consider how creationists might react to the scientific tree of life. Those in the "young Earth" sect of the doctrine believe that God brought various species into existence on days three through six of the week of October 23, 4004 BC, and because they think species cannot give rise to other species, these are all the species ever to have existed. Therefore, instead of offering us a single tree of life, the way scientists do, creationists would offer us what might be called a *forest of life*, with each species having its own tree. And because creationists think species cannot give rise to species, the trees in this forest will not branch out, the way the scientific tree of life does. They will instead resemble bamboo stalks, meaning that the creationist forest of life will resemble a bamboo grove (see Figure 4.7).

We have seen that as you go back in time, the scientific tree of life will tend to have fewer branches as you get closer to LUCA, the Last Universal Common Ancestor. This won't happen in the creationist forest of life, though. To the contrary, as you go back in time, the number of bamboo stalks will necessarily *increase*, as you include species that later went extinct. And at the bottom of the creationist forest of life, you will find stalks for all the species ever to exist. It will be quite a thicket.

Also, since people currently exist, no matter how far back in the creationist forest of life you go, they will exist. This means that the creationist forest of life will be compatible with the worldview expressed in the *Flintstones* cartoon show: man and dinosaur will be contemporaneous. Some creationists have supported this worldview by finding what

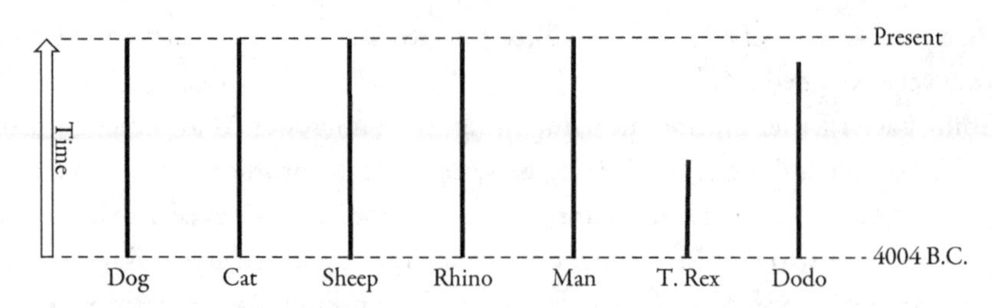

FIGURE 4.7. The creationist "forest of life." According to young Earth creationists, all the species ever to exist were created in 4004 BC. Consequently, species are represented not as branches of a tree, but as independent "bamboo stalks." Because *T. Rex* and the dodo bird went extinct, their "bamboo stalks" don't rise all the way to the present time.

they interpret as ancient stream beds—long since turned into stone—that have both dinosaur and human footprints.[11] Scientists tend to question either the authenticity of this evidence or the creationist interpretation of it.

Realize that it *is* possible to reconcile deeply held religious views with an acceptance of evolution. The Pope and Catholics in general seem to have done so.[12] If the religion that persecuted Galileo for saying that the earth moved can bridge the gap between science and religion, it ought to be possible for other religions to do so as well.

5

Your Sex Problem(s)

YOU ARE A SEXUAL BEING, in multiple senses of the word. Barring a chromosomal disorder, you possess either two X chromosomes, making you genetically a female, or one X and one Y chromosome, making you a male.[1] You are also probably a male or female, anatomically speaking. This, however, hasn't always been the case. In the first month of your development as a fetus, you would have been anatomically indistinguishable from a member of the opposite sex. Only then did your gonads receive a chemical signal instructing them to become either testes or ovaries, which in turn would determine the sex hormones to which you were subsequently exposed.

Besides being genetically and anatomically sexual, you are almost certainly a sexual being in the psychological sense of the word. This wasn't always the case, though. When you were four, you were psychologically asexual, but in your early teens, you became increasingly conscious of members of the opposite sex—or members of your own sex, if you were homosexual. And by the time you were in your mid-teens, you likely spent a considerable amount of time thinking about them and a fair amount of effort trying to get their attention. For better or worse, you had gained a sex drive.

During puberty, a male would likely have experienced his first orgasm. It would have been an astonishing event: how could an organ obviously designed for the convenient elimination of urine while standing be the source of such pleasure? And this question would quickly have been followed by another: can I make that happen again? From that point on, there would have been no turning back. The young man's priorities would have been dramatically reshuffled. A female going through puberty would not have been subject to anything as dramatic as an involuntary erection, but there would have been other indications of sexual arousal. Again, the discovery of sexual pleasure would have been a life-changing event.

To better appreciate the impact that gaining a sex drive has on one's life, consider what your life would have been like if you hadn't gone through puberty. The things that gave you pleasure as a child would not have been crowded out by the new pleasures you discovered, meaning that your childhood interests would probably have remained intact. You would not feel the urge to dress and act in a manner calculated to attract the interest of members of the opposite sex, nor would you feel the urge to create a family and do all the things necessary to maintain one. Your life would as a result have been vastly simpler.

Your sex drive has likely caused you to do some foolish and even dangerous things. It might also, on occasion, have brought you grief. You might have had your heart broken or contracted a venereal disease. You might also have seen your plans for the future turned upside down by an unwanted pregnancy, either in yourself or your lover. At the same time, though, your sexual nature has likely brought you both the pleasure of sexual activity and, I hope, the utter joy of experiencing reciprocated love. If you went on to become a parent, it also brought you the joys of parenthood. As a result, there probably aren't very many people who, if they had it to do over again, would choose an asexual existence.

FOR A LONG TIME AFTER LIFE AROSE, our planet was a sexless place. The cellular organisms that roamed it reproduced asexually, via cell fission. Sex had to wait for the evolution of eukaryotic organisms, about 2 billion years ago. The first eukaryotes would have continued to reproduce asexually, as some of them still do, but sometime in the next billion years, sexual reproduction became possible.[2] The reproduction in question, however, probably wouldn't have been very sexy, inasmuch as it involved no physical contact between the organisms that were reproducing. Organisms would simply have released their gametes into the environment. If by sheer chance a male gamete encountered a female gamete, fertilization would result.

This strategy is still used by plants. The male flowers of a ragweed plant might produce a billion grains of pollen. It is carried away by the wind, and if a pollen grain encounters a female ragweed flower, fertilization takes place and a seed will result. Many of the grains, however, don't find a female flower. They might instead end up in the sinuses of a hay-fever sufferer. But because ragweed plants make so many pollen grains, pollination is a viable survival strategy: all it takes for the ragweed population to hold its own is for a few of the grains released by each plant to find their target.

Some animals also use this gamete-dispersion strategy. When the moon is full, male corals release their sperm into the ocean, and females release their eggs. There is no evidence that they feel compelled to do this or that it feels good to do so. They are simply programmed to do it, the way you are programmed to have a heartbeat. The offspring that result from encounters between coral sperm and eggs will find themselves adrift, possibly far from their parents, where they might start a new colony.

At a more sexually sophisticated level, an animal will experience a sex drive. For this to be possible, the animal must be capable of a primitive kind of thinking. It must, to begin with, be able to feel good and bad, and it must be able to lay plans to increase the chance that in the future it will feel good and decrease the chance that it will feel bad. Otherwise, we would describe its behavior not as being "driven" but as being purely reflexive.

To better understand drives, consider your hunger drive. When mealtime is coming up, you feel unpleasantly hungry, or if you haven't eaten recently, you experience hunger pangs. In response to these feelings, you figure out how to get food. You experience pleasure while eating that food and feel pleasantly full when the meal is over. These feelings pretty much guarantee that during your life you will take the steps necessary to keep yourself nourished. Your sex drive works in a similar fashion. Go without sex for an extended period, and thoughts of it will start disrupting your thinking. Have sex, and you will enjoy the most intense (drug-free) pleasure a person can experience. It is an incentive system that keeps most people sexually active from puberty on.

Humans aren't alone in having a sex drive. Despite being simple little creatures, frogs also appear to have one. During mating season, a male frog will go to considerable lengths to find a female. On doing so, he will mount her and fend off other males who are trying to mount her. It isn't simply reflexive behavior; it is instead driven, albeit in a primitive manner.

WHEN THE FEMALE FROG RELEASES HER EGGS into the water, the male that has mounted her discharges his sperm, thereby fertilizing many of the eggs. Consequently, in the case of frogs, fertilization is external to both the male and female, and after the reproductive act, the eggs are on their own. Many will subsequently get eaten or will otherwise perish, but because so many of them are laid, there is a good chance that some of them will beat the odds and not only hatch but reach the age at which they can repeat the reproductive cycle.

External fertilization of eggs is clearly effective, but internal fertilization has one big advantage: eggs are more likely to get fertilized. Internal fertilization requires more from its participants, though. In particular, couples need to know how to have sex. Fortunately for them, they have a built-in sex instructor. If they simply pay attention to the signals their body is sending them and do the things that feel good, insemination will likely result.

Just because an egg is fertilized inside a female doesn't mean it has to develop there. Sea turtles mate at sea, with the male attaching himself to the shell of a female and then inserting his penis into her cloaca. She subsequently deposits the fertilized eggs into a hole she digs on a sandy beach and then abandons them. A chicken also lays her internally fertilized eggs, usually in a nest. She subsequently tends these eggs, though, keeping them warm and turning them, and after the eggs hatch, she continues to protect her chicks.

In mammals, eggs are not only fertilized within the bodies of females but develop there. The female's womb simultaneously keeps them safe, keeps them warm, and feeds them, making it an exceptionally good place for them to begin life. They emerge into the world not as eggs but as fully formed creatures. Consider, for example, whitetail deer. As a result of spending seven months in the womb, they are able to walk, albeit unsteadily, within minutes of being born.

There are, by the way, interesting variations on the internal-fertilization theme. Seahorses, for example, fertilize their eggs internally, but there is a twist. Instead of the male putting his sperm into the female, she carefully deposits her eggs into the brood pouch located on his belly, where he subsequently fertilizes them. He thereby becomes both the genetic father and the birth mother of his offspring.

YOUR SEXUAL DESIRES ARE FOCUSED. It is unlikely that you find yourself sexually attracted to a rock. It is also unlikely that you find yourself sexually attracted to, say, pigeons or bears. What instead attracts you is opposite-sex members of your species, but even within this group, you are picky. If you are a man, for example, you probably aren't sexually attracted to a seven-year-old girl or a seventy-seven-year-old woman. You will instead be drawn to, say, a twenty-seven-year-old woman—unless, that is, she has a carefully sculpted Van Dyke beard. For you, that will likely be a major turn-off. Similarly, if you are a woman, you will find yourself drawn not to just any male but to those who seem masculine. You might therefore find yourself attracted to a twenty-seven-year-old male with a muscular physique and Van Dyke beard—unless, of course, he also has an ample bosom.

People often find it difficult to explain their sexual preferences. This is because they don't get to choose them; they instead discover them, by realizing that people with certain characteristics have the power to arouse them sexually. These preferences are largely, but not entirely[3] a consequence of our evolutionary past. We are programmed to seek out, as sexual partners, individuals with whom we could successfully procreate. More precisely, we are (usually) programmed to be sexually attracted to members of the opposite sex who give evidence of being fertile. This is true even if the very last thing we want, at that point in our life, is to make a baby. This is presumably why we are turned off by bearded females or males with bosoms: it is evidence that they have hormonal problems that would likely result in infertility.

Think about life on the savannas of Africa a million years ago. Archaic humans who never experienced sexual desire would have failed to reproduce and would therefore have failed to become anyone's ancestor. The same can be said of those who were sexually attracted not to people but to pigeons or even worse, to bears; likewise for those who were sexually attracted to members of the same sex or to infertile members of the opposite sex. Individuals who were sexually attracted to fertile members of the opposite sex, though, likely reproduced, thereby giving rise to descendants who inherited the "wiring" responsible for their sexual preferences.[4] We are among those descendants, and we therefore share those preferences.

We have seen that one advantage of the internal fertilization and internal development of eggs is that they will emerge from their mother ready to function in the world. We humans are an exception to this. Despite having spent nine months in the womb, we aren't able to walk in a matter of minutes, the way deer are; it might instead take a year. And not only that, but during the first five or so years of our life, we will almost certainly perish without a parent to protect, feed, and care for us. This means that parents who were wired only to seek out and have sex with fertile-looking opposite-sex members of their species would be unlikely to have grandchildren, inasmuch as their children would be unlikely to survive to the point at which they could reproduce. The parents in question would therefore drop out of the evolutionary equation.

Consequently, besides being wired to seek out a member of the opposite sex who looks fertile, we are wired to look for a lover who gives evidence of being a good parent. Such a person will care about the wellbeing of any children that come into existence and take steps to ensure that they make it to adulthood. Whereas the former wiring causes us to experience lustful feelings, the latter can cause us to experience feelings of love, an emotion that, besides being directed toward offspring, can be directed toward other people—and in particular, toward our mate. And whereas feelings of lust typically last for a matter of minutes, feelings of love can last for decades.

ABOVE WE HAVE CONSIDERED THE COMPLICATIONS that arise in your life because of your sexual nature. This is only one of your sex problems, though. You will encounter an altogether different problem if you try to construct your hyperextended family tree. Each person on that tree will have two parents, one male and one female. These parents in turn will have two parents, and so on. We have seen the practical problems that this exponential growth rate will cause for anyone doing ancestral research. We have also pondered the family tree paradox: go back 2,000 years, and you will have more spaces on your tree to fill in than the total number of people ever to have existed. This paradox, as we have seen, dissolves when we realize that one person can appear in multiple places on a family tree.

Probe deeper into your ancestry, though, and you will discover a new problem. Your family tree assumes that each of your ancestors had two parents, one male and one female. We know, however, that there was a time before organisms reproduced sexually. This means that if you go back far enough on your family tree, you will arrive at ancestors who reproduced not sexually, by fusing an egg cell and sperm cell, but asexually, via cell fission. This in turn means that the binary structure of your family tree will break down. What will replace it, though? Something has to, or you wouldn't be here.

This problem becomes even more glaring if we approach it from the other side. Go back 3 billion years. As we have seen, at least one of the organisms that then existed was a direct ancestor of yours. As such, it would appear on your hyper-extended family tree. That organism would have been a single-celled microorganism that reproduced asexually, by fissioning into two "daughter" cells,[5] each of which was its clone. And therein lies the rub: how could this organism's descendants ever have transformed into sexual organisms? If one of these descendants, as the result of a mutation, switched from asexual to sexual reproduction, what would it mate with? By giving up the ability to reproduce asexually, though, it would lack the ability to clone itself. This sexual experiment would therefore have resulted in an organism that could not reproduce, meaning that the experiment would come to an end when it died.

This, in a nutshell, is your second sex problem: how could your ancestors have transitioned from asexual to sexual reproduction? This problem, like the family tree paradox, is less daunting than one might imagine. It seems problematic only because we assume that being sexual and being asexual are mutually exclusive states. We are encouraged to do so by our use of the term *asexual*. In English, following the practice in Greek, we use the prefix *a* to indicate that something is not the case: an atheist is not a theist, meaning that a person cannot be both an atheist and a theist. We are therefore led, linguistically, to conclude that it is impossible for an organism to be both sexual and asexual. But this turns out not to be the case.

IT IS POSSIBLE, TO BEGIN WITH, for members of a species to alternate between sexual and asexual reproduction. Yeasts, for example, are sexual organisms: if a yeast comes into contact with a member of the opposite sex, they fuse to make a "baby" yeast, containing the mixed DNA of the first two yeasts. When no member of the opposite sex is around, however, they have the ability to reproduce asexually, through a process known as *budding off*.

Aphids also have the ability to switch between sexual and asexual reproduction. In spring, eggs that have overwintered hatch to yield female aphids that spend their lives reproducing asexually. These females are parthenogenetic, meaning that they can make babies without the assistance of male aphids. Their offspring are clones of their mother and are therefore female as well.

In having these offspring, by the way, aphid moms don't lay eggs; they instead give birth to fully formed little aphids, ready to take on the world. And as moms go, they are remarkably prolific. In the month or so that they are alive, they might have a hundred babies. It is therefore theoretically possible for one aphid mom to have 1 million great-granddaughters. And not only that, but it is possible for the daughters that emerge from them already to be carrying offspring of their own.[6] In the 1970s, the sensationalist tabloid *National Enquirer* got readers' attention with the headline, "Baby Born Pregnant." In the case of aphids, this headline would not only be true but would be an unremarkable occurrence.

When they sense that winter is approaching, aphid moms switch gears and, without engaging in sex, start producing male aphids. The resulting aphids will *almost* be clones of their mother: they will have one fewer sex chromosome than she did.[7] These male aphids will subsequently impregnate females—at last, sexual reproduction—which in turn will produce not living baby aphids but eggs that overwinter. When spring comes, they hatch as females, and the cycle begins again.

And while we are on the topic of aphids, one more thing. Some aphid species have a symbiotic relationship with ants. The ants are like dairy farmers who care for their cows not so they can eat them but so they can drink the milk they produce. These ants protect their aphids from predators, such as ladybugs, and in turn the aphids, when stroked, provide their ant guardians with a drop of a sweet liquid called *honeydew*. Having learned all this about aphids, one is tempted to grow rose bushes, not for the roses they will produce but for the aphids they will attract. To have such wondrous creatures in one's garden!

Once we give up the assumption that being sexual and being asexual are mutually exclusive states, either in a species or in a member of a species, we can explain how it was possible for your ancestors to have transitioned from asexual to sexual reproduction. There would have been a time when all of your ancestors were asexual, a time when they were all sexual, and a time, in between these two periods, when they were capable of reproducing either asexually or sexually, depending on their circumstances. During this transitional period, your family tree will resemble that of an aphid, inasmuch as your ancestors would have engaged in a mix of sexual and asexual reproduction (see Figure 5.1).

SO FAR, WE HAVE CONSIDERED TWO of your sex problems. The first involves the challenges you face as a result of having a sex drive. The second would be encountered if you attempted to construct your extended family tree: how would you handle the transition from asexual to sexual ancestors? This brings us to a third sex problem that you, personally, might not find too worrisome—unless, that is, you are a biologist. The problem in question arises when we attempt to explain why your ancestors would have abandoned asexual reproduction in favor of sexual reproduction.

Sexual reproduction has many disadvantages. In sexual reproduction, two cells fuse to become one, thereby halving the cellular population; in asexual reproduction, one cell fissions to become two, thereby doubling the cellular population. Furthermore, to reproduce sexually, you have to take the time and expend the energy necessary to find a mate, and if you find one, you run the risk of contracting a venereal disease if you mate with them. You would therefore think that sexually reproducing cells would find it difficult to compete against their asexual rivals in the battle for resources. While sexual organisms were going through all the bother of finding a mate in order to make a single offspring, their asexual rivals would be happily doubling and then redoubling their population.

One other problem with sexual reproduction is that it can undo the competitive advantages that genetic mutations provide. Suppose that an organism, as the result of

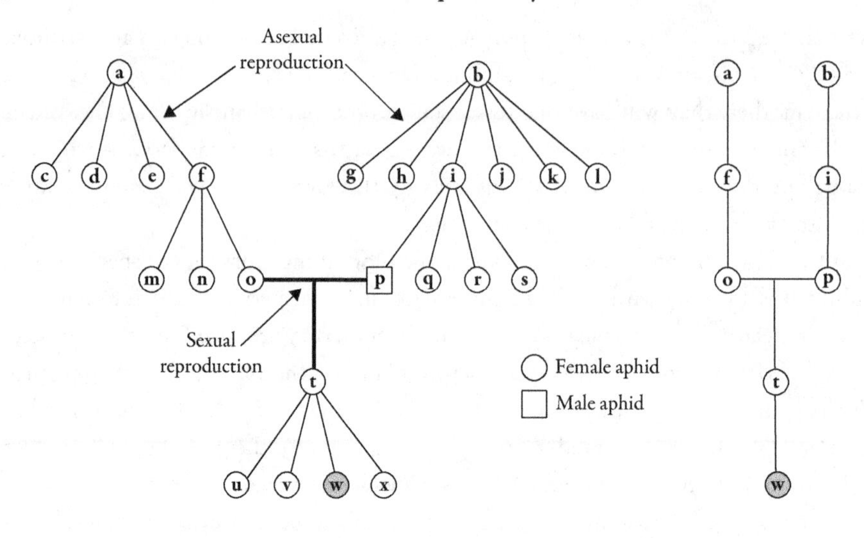

FIGURE 5.1. On the left is a much simplified tree of descendants for the aphids labeled *a* and *b*. In this tree, female and male aphids are indicated with circles and squares, respectively. The tree depicts the mix of sexual and asexual reproduction employed by aphids. On the right is the family tree for the aphid labeled *w* (shaded). Your own hyperextended family tree would at some point resemble that of the aphid, inasmuch as it would show a mix of sexual and asexual reproduction.

such mutations, ends up with the ideal combination of genes for the environment in which it finds itself. An asexual organism would be able to pass on those exact genes, meaning that its cloned descendants would have a significant advantage over their rivals. A sexual organism, by way of contrast, would be able to pass only half of those ideal genes to its offspring; the other half would be provided by its genetically less-than-ideal mate. What a waste!

That said, it is clear that sexual reproduction must confer significant benefits on those species that practice it; otherwise, it wouldn't be as common as it is. Virtually all eukaryotes are sexual, at least some of the time, and the vast majority can reproduce only by means of sex.[8] But what exactly, biologists wonder, are the benefits of sexual reproduction?

The standard answer to this question is that sexual reproduction involves a shuffling of genes that leads to greater genetic diversity among members of a species. This diversity allows sexual organisms to "experiment" in order to find the set of genes that works best in the environment in which they find themselves. Yes, there will be setbacks when an organism with the ideal combination of genes mates with an organism that lacks it, but the process of becoming well-suited to the environment will nevertheless be faster than simply waiting for mutations to produce the right combination. Furthermore, when the environment changes, a species with genetically diverse members will be more likely to survive: chances are that *some* members of that species will be adequately suited to the new environment. Their survival will result in the survival of their species—not that they will realize as much.

Although this is, as I have said, the standard solution to the paradox involving sexual reproduction, many biologists find it less than satisfactory, and so the debate over why sex exists continues in scientific literature.[9]

ONCE WE CONVINCE OURSELVES that sexual reproduction is, evolutionarily speaking, a good idea, it isn't difficult to think of ways to make it even better. Here are some of them.

Why not, on gaining the ability to reproduce sexually, retain the ability to do so asexually? Our distant ancestors, as we have seen, had both abilities but subsequently relinquished their ability to reproduce asexually. Had we retained both reproductive abilities, we would have been able to enjoy the benefits (whatever they may be) of sexual reproduction but would also have been able to reproduce, if circumstances called for it, without having to find a sexual partner. In particular, if we were lucky enough to have genes that were ideally suited to the environment in which we found ourselves, we wouldn't have to dilute those genes by engaging in sexual reproduction.

Why not allow organisms capable of sexual reproduction to change from one sex to the other, depending on circumstances? This may sound farfetched, but it is an ability that bluehead wrasse fish possess. When the dominant male of a population dies or disappears, the largest female starts acting like a male and then changes coloration to look like one. In the final step of this transformation, its ovaries turn into testes.[10] Clown fish exhibit similar behavior, but in the opposite direction: when the dominant female dies, the dominant male will change sex to take her place.[11]

Why not allow organisms to be simultaneously male and female? When a species is divided into males and females, as our species is, we can mate with only half the members of our species. If a species is hermaphroditic, the number of possible sex partners doubles: members can mate with any other member, or for that matter, with themselves. Wouldn't that be grand?

If a species is going to be sexual, why stop at two sexes? If there are two (and a species isn't hermaphroditic), each member can mate with only half of the members of it species; if there were three, it could mate with two-thirds of them; and if there were ten, it could mate with nine-tenths of them. Seen in this light, we are forced to conclude, as does biochemist Nick Lane, that having two sexes represents "the worst of all possible worlds."[12]

The reader might respond to this suggestion by saying that although having more than two sexes might in theory be advantageous, it is a practical impossibility. Sex, after all, is a binary thing: a sexual organism must be either male or female, or if it is hermaphroditic, both male and female. Realize, though, that many multiple-sex species exist, including *Tetrahymena thermophila*. This protozoan has seven sexes, Types I through VII, and can mate with an individual of any sex but its own. Or at least when it is sexual it has seven sexes; it also goes through a phase during which it reproduces asexually. And when it switches to sexual reproduction, which sex it assumes is random.[13] *Physarum polycephalum*, a slime mold, has an even more exotic sex life, inasmuch as it can be one of at least 13 sexes. And not to be forgotten is *Schizophyllum commune*, a mushroom with 28,000 different sexes.[14]

Examples like those cited above have led some to compare the "mindless" evolutionary process to the process used by an indefatigable tinkerer.[15] Evolution is perfectly willing to take an organism that basically works and try to make it work even better—if not in the current environment, then in one that might arise in the future. Two sexes work? Then what about 7, 13, or even 28,000? Such tinkering is largely responsible for the vast diversity of life forms that have arisen on our planet, for their ability to endure the catastrophes our planet has experienced, and for the remarkable extent to which they have taken possession of the earth's land, oceans, and atmosphere.

6

The Neanderthal in Your Family Tree

IN 2010, RESEARCHERS SUCCEEDED in recovering the DNA of Neanderthals who lived nearly 40,000 years ago.[1] Subsequent analysis revealed that some of the genes in that DNA had found their way into modern humans. The only way this could happen, though, is if sometime in the past, Neanderthals mated with our *Homo sapiens* ancestors, those encounters resulted in offspring who themselves had offspring, and so on, down to the present time. This in turn means that if most of us did our extended family tree, we would find many Neanderthals lurking there.[2]

Some people were disturbed to learn of their Neanderthal ancestry. This was in part because they regarded *Homo neanderthalensis* as an inferior species whose members were little more than grunting, brutish cavemen. But such indignation was clearly inappropriate. For one thing, Neanderthals were not stupid and clumsy. They were skilled hunters who could make and use sophisticated weapons and tools. They could control fire, make crude clothing, and construct dwellings. They had complex social groups and probably could use language. They also seem to have, in at least some cases, ritually buried their dead. They were clearly sophisticated hominins. Furthermore, it seems strange to object to having Neanderthals as ancestors, given that we also have worms as ancestors.

People who could remember their high school biology, though, found the Neanderthal announcement disturbing for another reason. They had been taught, first of all, that Neanderthals and *Homo sapiens* belonged to different species, and second, that members of different species cannot mate and have fertile offspring.[3] The presence of Neanderthals in their family tree meant that one of these doctrines had to be mistaken, but which?

In this chapter, we will explore our Neanderthal ancestry and in the process of doing so reexamine the concept of a species. We took this concept for granted when we examined the tree of life in chapter 4. It turns out, though, that species are not the clear-cut categories one might imagine. They are inherently vague, and it is because of this vagueness that Neanderthals can appear on our family tree.

SUPPOSE YOU GAINED THE ABILITY to travel back in time. It would be a wonderful way to do ancestral research. You could not only meet your ancestors but get inside information about your family tree: your great grandmother might tell you who, contrary to published records, your *real* great grandfather was. And after you had gotten this information, you could thank your ancestors—or who knows, maybe blame them?—for the role they played in making your life possible. You could also use your trip to do some serious scientific research. As you went back in time, you could take a census of the species that then existed and determine how they were related to the species you had encountered on your previous stop.[4] You could then use this information to construct a complete and accurate tree of life.

Be forewarned, though, that time travel—assuming that such a thing is possible—has its hazards. There is, to begin with, a danger that you will inadvertently do something to disrupt the chain of events that eventually leads to your existence. It is a danger that has long fascinated philosophers and science-fiction writers. And even if you are careful to avoid these "existential dangers," there are other very real dangers connected with time travel.[5]

I live in Dayton, Ohio. If I went back in time 20,000 years, I would, when my time-travel machine stopped, find myself under 1,000 feet of ice. It was, after all, the peak of the most recent ice age, and the ice in question covered most of Ohio. It is responsible for the moraines that can be found in my part of the state. These long hills of rocky soil at one time marked the edges of glaciers. Travel back 400 million years, though, and I would find myself not under ice but under the tropical sea responsible for the crinoid fossils I have found near my home. Travel back 650 million years, and I would again find myself under ice that, according to the Snowball Earth hypothesis, covered virtually the entire globe.

Suppose that I instead started my trip in New York City. Traveling back 20,000 years, I would also find myself under a glacier. The evidence for this is the glacial striations on the bedrock outcroppings in Central Park. Travel back 400 million years, though, and I would find myself not under the tropical sea that submerged Dayton but atop the mountain range—or maybe under it, depending on how time travel works—that is thought to have been 15 times higher than any of Manhattan's skyscrapers.[6] It is these mountains, by the way, that are responsible for the bedrock necessary to support those skyscrapers. One upside of starting my time travel in New York City is that it would be possible, 300 million years ago, to travel to Africa by taking a short walk over dry land.

This is because at that time, all the earth's continents were pushed together into the supercontinent known as *Pangaea*.

And speaking of mountains, if I started my time travels atop Mount Everest, I would find myself at a lower elevation on each stop, until finally, 400 million years back, I would find myself at the bottom of the sea responsible for the crinoid fossils that can be found on Everest's peak. If I instead started my travels at the top of the Matterhorn in Switzerland, I would, 100 million years back, find myself in Africa. This is because the very tip of the Matterhorn was originally part of the African tectonic plate.[7]

And no matter where I started my time travels, I would, as I went back in time, expose myself to ancient microorganisms. Because my body had not had an opportunity to develop immunity to them, they might kill me. Furthermore, if I traveled back more than a billion years, I would find it difficult to breathe. This is because until that time, most of the oxygen produced by photosynthetic organisms would have been taken up in chemical reactions, such as combining with iron to make rust, rather than floating free in the atmosphere.[8]

BUT ENOUGH ABOUT THE DANGERS of time travel. Suppose you found a way to avoid them, perhaps by traveling back in an unsinkable, crush-proof, insulated capsule with its own air supply. Suppose that to be thorough, you decided to make stops every 100,000 years and that on each of these stops you would roam the world to see what organisms you could find.

In some cases, the organisms would be essentially the same as you encountered on your previous stop. Gingko trees, for example, have existed in their current form for more than 100 million years.[9] Most of the organisms you encountered, though, would be somewhat different than they formerly were, but with some DNA detective work, you could become quite confident about which species were the ancestors of which later species.

As you went back in time, you would notice a curious phenomenon: certain pairs of species would increasingly resemble each other. In your first few time travel stops, for example, humans would have looked quite different from chimpanzees. As you went back further, though, your human ancestors would look more and more like the ancestors of chimpanzees, until finally you couldn't distinguish between them: two species would have merged into one. Something similar would happen with horses and zebras, with lions and tigers, with blue whales and hippos, and even more astonishingly, if you went very far back in time, with blue whales and mosquitos. This last common ancestor, by the way, would look like neither a whale nor a mosquito.

This "species merger" phenomenon is an illusion created by traveling back in time. What is really happening is that one species is dividing into two. As a result of this phenomenon, it would generally be the case that the further back in time you went, the fewer species there would be to keep track of; indeed, this has to be the case for all currently existing species to have descended from a single living organism, as evolutionary biology says they did. Realizing this would give you an incentive to press on in your time-travel research. We saw that constructing your family tree is a disheartening enterprise, since each ancestor you add gives rise to two more ancestors to research. Constructing the tree of life, by way of contrast, becomes easier the farther back you go, since there will be ever-fewer species to deal with.

But your desire to press on with your research will be tempered by your discovery of another curious phenomenon. On each time-travel stop, besides finding species that were slightly different from those you had most recently encountered, you would find species that were radically different from any you had seen. At first, the sudden appearance of these species might surprise you, but then you would realize that this is yet another illusion created by time travel. These species aren't suddenly popping into existence; they are simply the species that had died out between this stop and your previous stop. This is why on your first stop, 100,000 years back in time, you would encounter, in the flesh, species that you had only heard about when you lived in "the present." Among them would be woolly mammoths and saber-toothed tigers.

On some stops, the increase in the number of "new species" would be particularly dramatic. On your stops before 66 million years ago, for example, you would have seen nary a dinosaur,[10] but then, on your next stop after that, they would be abundant. This is because the asteroid that took out the dinosaurs would have hit during the interval between this stop and your previous stop. You would likely be irritated by this increase in your research workload, and the situation would become dramatically worse when, about 252 million years ago, your travels took you past the time of the End-Permian extinction event. This event is estimated to have taken out 96 percent of species, meaning that there would be a sudden 25-fold increase in your workload. Groan!

This might, for you, be the straw that breaks the camel's back: you might decide to give up your research and go back to the future. If you were to press on, though, you would find that the trend to fewer species would continue, as the branches of the tree of life "merged." Finally, you would reach LUCA, the Last Universal Common Ancestor of all currently living things—which, by the way, would neither have been the first living organism on our planet nor the only organism to exist at that time. In chapter 8, we will see why.

And what, one wonders, would this organism have been like? Inasmuch as it existed in a time before multicellular organisms had arisen, LUCA would have been a single-celled organism. And since LUCA was by definition the ancestor of all currently living things, by looking for similarities in the genomes of a wide range of living things, biologists can gain insight into LUCA's genome and thereby gain insight into what sort of organism it was. Such research suggests that LUCA was most likely a heat-loving anaerobe that fed on hydrogen and that it probably lived in a thermal vent on the ocean floor.[11]

Suppose that after tracking down LUCA, you decided to return to the present. The scientists who greeted you would congratulate you for having filled in the tree of life, but they might then express their disappointment that you didn't press on to find out how LUCA came to exist and, more important, how inanimate matter transformed into living things. We will return to this question in chapter 8. In the remainder of this chapter, though, let us use the time travel story as a device for exploring the concept of a species.

YOUR FAMILY TREE, AS WE HAVE SEEN, takes people as its unit of investigation and shows how they are related to each other. The tree of life, by way of contrast, takes species as its unit of investigation and shows what species gave rise to what other species. In its classic form, the tree of life takes species to be discrete things. In its depiction of which species gave rise to which, it therefore shows one species turning into another, with no transition (see Figure 6.1). But this is not how things work, as you would see if you traveled back in time—not in 100,000-year jumps, in order to determine which species gave rise to which other species so you could construct the tree of life, but in much shorter jumps, in order to identify the parents of each of your ancestors so you could construct your extended family tree. You might start your journey by traveling back to visit your eight great-grandparents, then go further back to meet their parents, and so on.

You would notice, as you went back, that your ancestors were changing. At first, the changes would be superficial—they would, for example, do different things with their hair and dress differently than you do. But when you compared your ancestors of 100,000 years ago with yourself, you would notice other, more fundamental changes. The structure of their faces and shape of their heads would be different than yours, although still recognizably human. Your ancestors of 1 million years ago would have faces and heads that you might describe as being weirdly human. And your ancestors

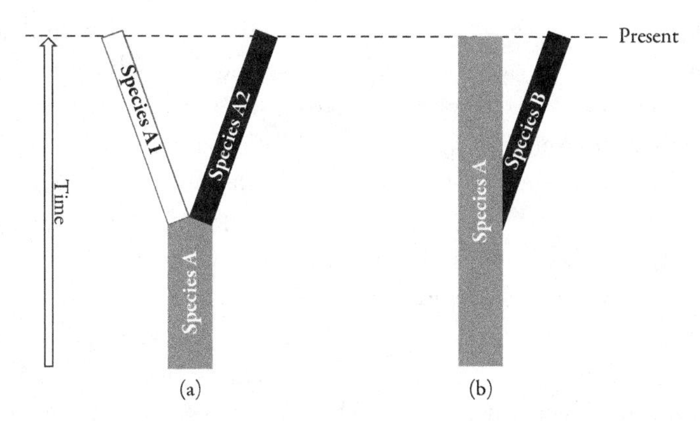

FIGURE 6.1. In the classic form of the tree of life, species are depicted as coming into existence "instantaneously." In some cases, one species transforms into two, as is shown in tree (a). In other cases, one species buds off from another, while the original species continues to exist, as is shown in tree (b).

of 4 million years ago wouldn't just have different faces and heads; their bodily proportions, along with their posture and gait, would be quite unlike those of today's humans. On seeing an ancestor from this period walk toward you in the dusk, you would probably react by thinking not that *someone* was coming, but that *something* was.

Although this transformation of your ancestors would be profound, it would have taken place in imperceptible increments. Consequently, there would never be a "dividing line" generation that you could, with any justification, point to as the end of one species and the beginning of another. Members of one species don't have offspring of a different species![12]

Biologists sort currently existing life forms into species on the basis of their physical characteristics, their physiology, their behavior, their body chemistry, and in recent decades, their genomes. Paleontologists have less information to work with when they sort extinct life forms into species on the basis of their fossilized remains. They also, in most cases, have incomplete fossil records to work with. But this last factor in some ways simplifies the sorting job. For example, when Dutch paleoanthropologist Eugène Dubois found a curiously shaped partial skull in 1891, he did not hesitate to assert that it was evidence for the existence of a new species, located somewhere between man and apes on the tree of life. That species came to be known as *Homo erectus*, and the fossil Dubois found is regarded as the benchmark fossil for that species.

If paleontologists had more fossil evidence available to them, though, the sorting problem would become much more difficult. To see why I say this, suppose that while doing the time travel in conjunction with researching your family tree, you decided to do paleontologists a favor: for every 10,000 years back you traveled, you would collect a complete skeleton of a direct ancestor of yours who had passed away shortly before your visit. Suppose you ended up bringing back 700 very well-preserved skeletons, with the oldest dating back 7 million years, the point at which our ancestors parted company with the chimpanzees.

For paleontologists, access to these skeletons would be a good news/bad news situation. The good news is that this access would give them an incredibly deep understanding of human evolution. The bad news is that it would seriously complicate their efforts to sort our hominid ancestors into various species. The skeleton collection would make it abundantly clear that species are not discrete things—that one species instead slowly and subtly transmutes into another. So what justification could there be for claiming, say, that skeleton #19, from 190,000 years ago, was a member of the species *Homo sapiens*, while skeleton #20, from 200,000 years ago, was not?

This same issue arises if we try to assign names to colors. If we are presented with the complete spectrum of colors, with shades of red gradually changing into orange, then yellow, green, blue, indigo, and violet, it will be difficult to say, for example, which exact shade of red should be our standard for redness and which exact shade of green should be our standard for greenness. Any choice we make will seem arbitrary. Present us with a gappy color spectrum, though, in which large sections are blacked out and little sections

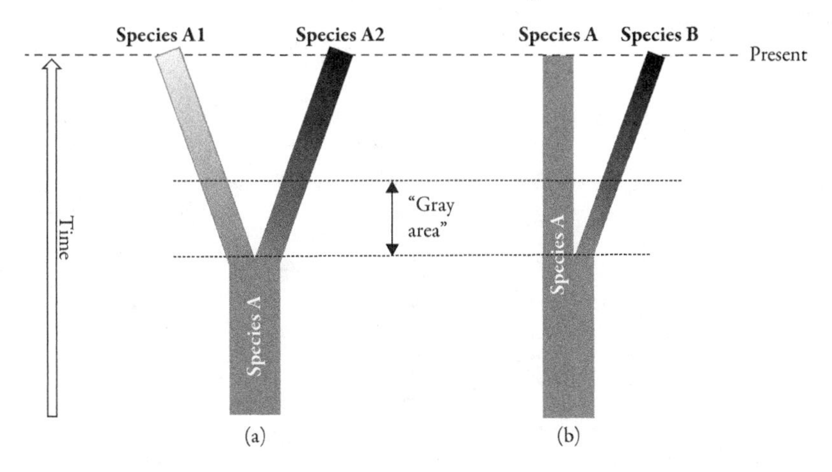

FIGURE 6.2. In the improved tree of life, species are depicted as coming into existence gradually. Tree (a) shows one species gradually transforming into two. Tree (b) shows one species continuing to exist as another species gradually buds off from it. In either case, there will be a "gray area" interval, during which it will be unclear which species an organism belongs to.

of color are visible,[13] and the job of naming the colors becomes easy. There might be only two reddish portions visible, so it won't be much trouble to pick one of them out as being "most red." It is because the fossil record is similarly gappy that paleontologists feel confident in identifying ancient species.

IN LIGHT OF THIS DISCUSSION, LET US REVISIT the tree of life. In its classic form, as I've said, this tree takes species to be discrete things, meaning that new species appear instantaneously (see Figure 6.1), but this assumption, as we have seen, is unrealistic. A more accurate tree of life will depict species as gradually transforming into other species, as is shown in Figure 6.2.[14] During the transformation process, there will be a "gray area" interval, during which it will be difficult to distinguish between members of the two species.

Let us also pause here to reexamine the concept of extinction. When the members of a species no longer roam the earth, we say that the species is extinct. Consequently, *Tyrannosaurus rex* and *Homo heidelbergensis* count as extinct species. Realize, though, that there are two ways in which the members of a species can cease to exist (see Figure 6.3). One is if every member of that species dies without leaving any descendants. This was the fate of *T. rex*. The other is if a species transforms, through evolutionary processes, into another species. Although the former species will no longer exist, the descendants of members of that species will. This is the fate of *Homo heidelbergensis*; indeed, you are one of those descendants.[15]

There will doubtless come a time when we humans no longer roam the earth. The only question is what will cause us to disappear. Given our talent for driving species into

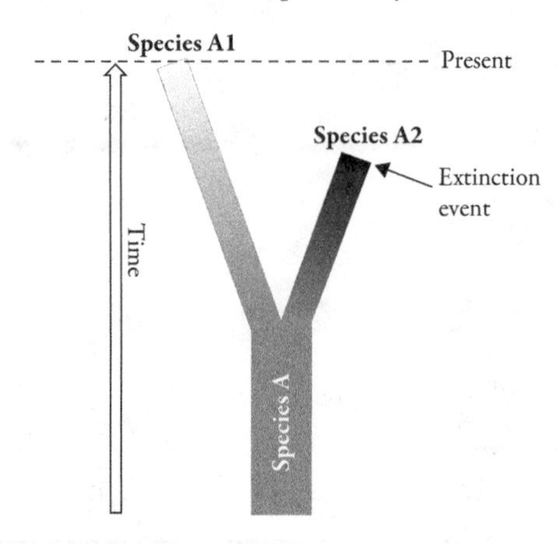

FIGURE 6.3. There are two ways in which a species can cease to exist. One is for all the members of that species to perish, which is what happened to species A2. The other is for the descendants of the members of a species to transform, via evolutionary processes, into members of a different species, which is what happened to species A.

extinction, it is possible—some would say likely—that we will drive ourselves into extinction, perhaps as the result of nuclear or biological warfare. Alternatively, there is nothing to prevent us from going out the way the dinosaurs did, as the result of an asteroid strike. Even if we avoid this fate, though, it is virtually certain that our species, if it exists long enough—or is subjected to enough genetic engineering—will transform into a different species. And ironically, if our descendants, hundreds of thousands of years from now, were to dig up and analyze the bones we twenty-first-century humans leave behind, they might conclude that we didn't really belong to a species—that we instead represented an intermediate stage between *Homo heidelbergensis* and the next (yet to have evolved) species. Oh, the indignity!

For further insight into the way one species transforms into another, consider what might be called the *anthropocentric tree of life*. In the usual tree of life, our species appears as one of many branches. In the anthropocentric tree, we place our species and its evolutionary ancestors along a single straight branch. We also use shading like that used in Figure 6.2 to make it clear that species transformations are gradual processes. If we choose to include other species on our tree, they appear as offshoots from our main branch, with the offshoots in question emerging from the species that we and they have in common. For a vastly simplified version of such a tree, see Figure 6.4.

This tree, you might be thinking, resembles the one given by Ernst Haeckel (see Figure 4.5). It makes it look like man represents the inevitable destination of the evolutionary process, with the other species being mere side trips. Realize, though, that a similar tree could be created for any other species. I am presenting this tree rather than that of, say, *Halyomorpha halys*—also known as the brown marmorated stink bug—because this is the tree of *your* species and is therefore likely to be of more interest to you than the *H. halys* tree would be.

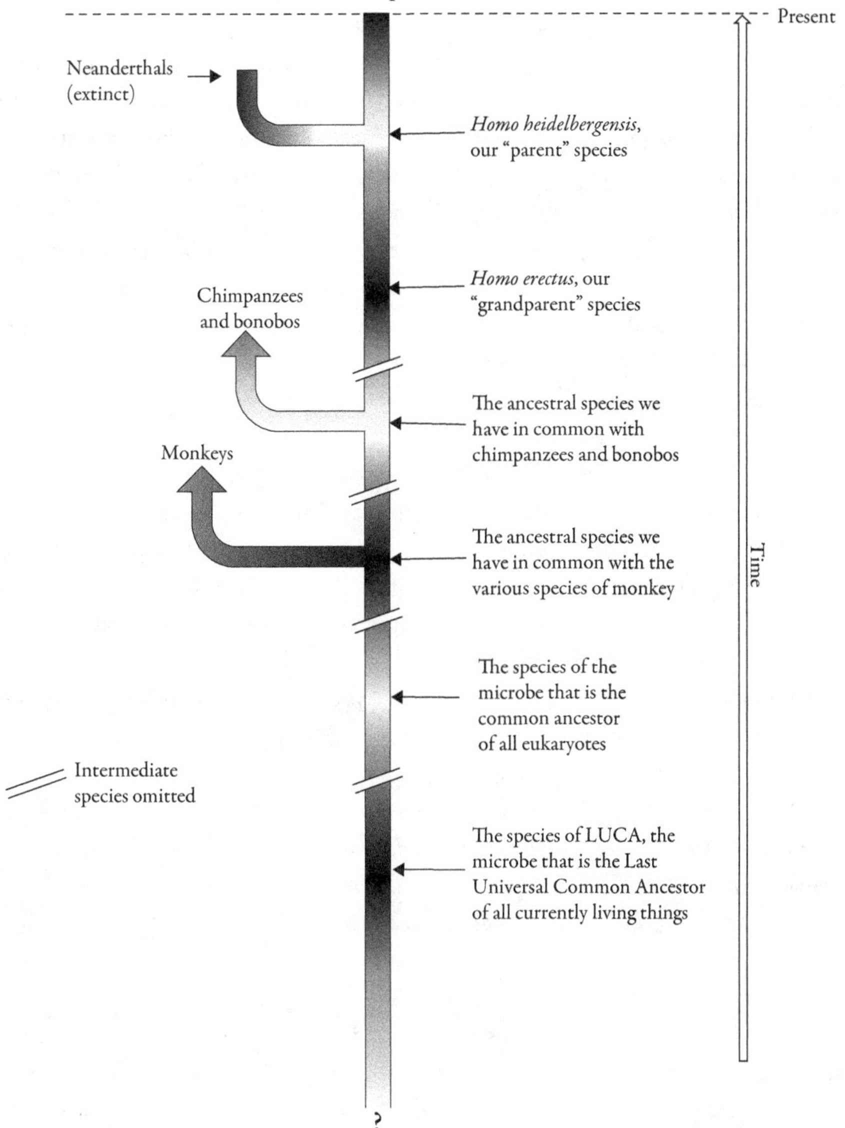

FIGURE 6.4. The anthropocentric tree of life, vastly simplified, in which the ancestry of our species is represented by a straight branch, and other currently existing species are represented by offshoots from that branch. The gradual change in darkness indicates the gradual transformation of one species into another. The vertical time axis is not drawn to scale. Also, very many species branch offs are omitted.

ONE WAY FOR A NEW SPECIES to come into existence is for the members of a species to be physically separated. Suppose, for example, that one group of monkeys has "rafted" to an island[16] while another group has remained on the mainland. At first, there will be a high degree of reproductive compatibility between the islanders and the mainlanders. As time passes, though, and the groups drift genetically, there will be fewer and fewer intergroup matings that are reproductively compatible: if we reunite island monkeys with mainland monkeys, some matings will succeed while others will fail. Finally, there will come a time at which none of these matings will succeed. Realize that this change in reproductive compatibility, like the transformation from one species to another, will not happen in one generation and probably not even in a hundred. It will instead take place very gradually, over thousands of generations.

Languages, by the way, evolve in much the same manner as species do. Take a group of people who have a language in common, divide them into separate groups, and cut off communication between them. With the passage of time, the way they use words will change. Along these lines, think about the way British English has, as the result of a few centuries of separation of its speakers, transformed into American English and Australian English. Likewise, think about the way Vulgar Latin has, as the result of many centuries of separation of its speakers, transformed into dozens of Romance languages. Consider, in particular, the following sentence in Vulgar Latin: *Ea claudit semper illa fenestra antequam de cenare,* which means "she always closes the window before she dines." Below is that same sentence in various other Romance languages.[17]

Spanish: *Ella siempre cierra la ventana antes de cenar.*
French: *Elle ferme toujours la fenêtre avant de dîner.*
Italian: *Ella chiude sempre la finestra prima di cenare.*
Aromanian: *Ea ãncljidi totna fireastra ninti di tsinã.*

In the same way as languages drift apart when their speakers are separated, genomes drift apart when members of a species are separated. And in the same way as speakers of a language, if separated for a long enough time, will no longer be able to communicate with each other, members of a species, if separated for hundreds of thousands of years, will no longer be reproductively compatible. And one other thing: in much the same way as we can use trees of life to show how species are related, we can use linguistic trees to show what languages evolved from what other languages.

WE GENERALLY TREAT REPRODUCTIVE COMPATIBILITY as if it were a dichotomous property that two individuals either have or lack, but things are in fact more complicated than this. Allow me to explain.

Depending on your genes, you have one of four blood types, A, B, O, or AB.[18] Your blood will also be either Rh-negative or Rh-positive. Suppose that Nancy, Nick, and Paul are three healthy 20-year-olds, that Nancy and Nick—the ones whose names start with *N*, for *negative*—have Rh-negative blood, and that Paul has Rh-positive blood. There is a very good chance that if Nancy and Nick mated, they would have many healthy babies. If Nancy instead mated with Paul, the resulting baby would probably be Rh-positive[19] and would probably develop without complications, but there is a wrinkle. The presence of an Rh-positive baby in Nancy's womb would cause her immune system to make antibodies against the Rh factor. If a baby that Nancy and Paul subsequently conceived was also Rh-positive, Nancy's primed immune system would attack it. It is therefore unlikely that it would, without medical intervention,[20] be viable.

So is Nancy fertile or infertile? It depends on whom she mates with and when they mate. She and Nick are reproductively compatible: if they mate, healthy offspring will almost certainly result. She and Paul are initially reproductively compatible but thereafter are likely to be reproductively incompatible. Nancy can therefore best be described as being neither fertile nor infertile; she is instead reproductively compatible *with some people, some of the time*. The same can be said of Paul. He won't be universally reproductively compatible, in the sense of being reproductively compatible with every woman. He won't, in particular, be reproductively compatible with Nancy after their first Rh-positive baby—unless there is medical intervention. Meanwhile, Nick, as a male with Rh-negative blood, will be more reproductively compatible than Paul is. His Rh-negative status will be a problem with neither Rh-negative nor Rh-positive women.

Fertility is similarly complex in other species. Horses and donkeys, for example, are usually reproductively compatible. Their offspring will be a mule, if the horse is female and the donkey is male.[21] This mule, however, will be a sterile hybrid, meaning that it will not be reproductively compatible with either horses, donkeys, or mules—usually. Occasionally, a mule will mate with a donkey and produce offspring,[22] and this kind of reproductive compatibility is by no means unique: about a hundred known mammal species can interbreed and produce fertile offspring.[23]

It was a genetic issue that jeopardized the reproductive compatibility of Nancy and Paul. In other cases, a couple might be reproductively incompatible because the woman's immune system attacks the man's sperm before they can make it to her egg. There are also cases in which a sperm makes it to the egg, but the chromosomes it carries are unable to properly pair with those of the egg. This will be the case if the sperm and egg have a different number of chromosomes—which is one of the reasons we humans, with our 23 pairs of chromosomes, are reproductively incompatible with chimpanzees, with their 24.

THIS BRINGS US BACK to the Neanderthals on your family tree. *Homo sapiens* and Neanderthals are both descended from *Homo heidelbergensis*, but Neanderthals evolved in Europe, hundreds of thousands of years before *Homo sapiens* evolved in Africa. As a result, the two groups drifted in different genetic directions, meaning that by the time our ancestors left Africa and encountered Neanderthals, the number of reproductively compatible parings between members of the two groups would likely have declined. Many matings would not yield offspring, and when offspring did result, many of them would not have been fertile. Every now and then, though, matings would, by the luck of the draw, have resulted in a hybrid offspring who was reproductively compatible with members of the species *Homo sapiens*. One or more of these offspring must have had descendants, who in turn had descendants, and so on, down to those currently existing humans who have Neanderthal DNA.

Whether or not you have Neanderthal DNA in your genome depends on where your ancestors went and whom they had sex with. If they stayed in Africa—as did the San people who today inhabit the Kalahari Desert—you will have little Neanderthal DNA. (If you are looking for "purebred" members of our species, the Kalahari Desert is a good place to find them.) But if your *Homo sapiens* ancestors left Africa and headed toward Europe or Asia, they would have encountered Neanderthals and might subsequently have mated with them. This is how most modern humans who are ethnically European or Asian came to have between 1 and 4 percent Neanderthal DNA.[24]

On leaving Africa, our ancestors also would have encountered the Denisovans, another extinct hominid species, the DNA of which has been obtained and analyzed.[25] Matings between *Homo sapiens* and Denisovans must also have taken place,[26] but such encounters were either less common than encounters with Neanderthals or were less likely to result in fertile offspring. This would explain why Denisovan DNA is, with some interesting exceptions,[27] less common in modern humans than Neanderthal DNA. And in thinking about the sources of our DNA, we should keep in mind that besides carrying the DNA of Neanderthals and Denisovans, we carry DNA from the species *Homo heidelbergensis* and *Homo erectus*, who were our parent and grandparent species, respectively. We also carry DNA from the ancestor we have in common with chimpanzees.

Earlier in this chapter, I mentioned that some people were disturbed to hear that Neanderthals lurked in their ancestry: they thought Neanderthals were subhuman. A case can be made, though, that rather than being ashamed that there are Neanderthals in our family tree, we should be grateful for their presence. Because they moved to Europe and Eurasia long before *Homo sapiens* did, they were better adapted to those regions than the *Homo sapiens* newcomers. In particular, the Neanderthals could better withstand cold. By mating with them, our *Homo sapiens* ancestors could quickly acquire the genes that enabled us to inhabit Europe and much of Asia.[28] And the modern-day inhabitants of the Tibetan plateau, besides acquiring cold-tolerance genes from their Neanderthal ancestors, apparently acquired altitude-tolerance genes from their Denisovan ancestors.[29] Lucky them! Indeed, had our human ancestors not acquired genes from these other species on their way out of Africa, it is unlikely that their subsequent existence outside of Africa would have been as successful as it was.[30]

Before we end this discussion of your Neanderthal ancestry and the concept of a species, let me describe one more variant of the tree of life. We have seen that in their classic form, trees of life are misleading, inasmuch as they show one species instantaneously giving rise

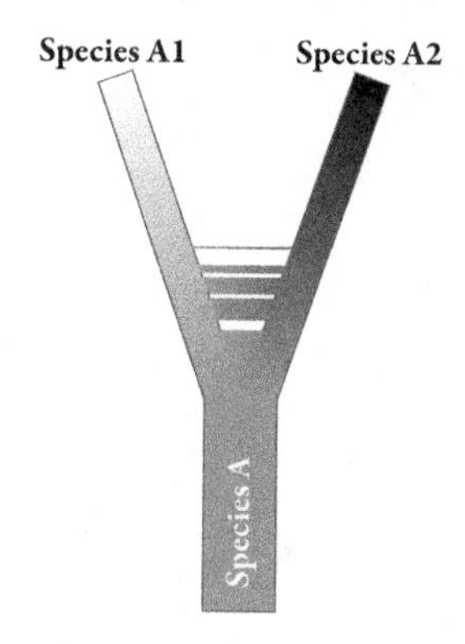

FIGURE 6.5. In a "webbed" tree of life, interbreeding between closely related species is indicated by horizontal "twigs" between the tree's branches. As the species diverge, the reproductive compatibility between members of the two groups declines, and interbreeding becomes increasingly rare.

to another. We have also explored an improved tree of life in which species slowly evolve into new species. The discussion presented here suggests that if we want an even more realistic tree of life, we need to modify our improved tree by allowing "webbing" between its branches, to indicate interbreeding between members of closely related species (see Figure 6.5). It is less tidy than our previous trees but does a better job of depicting the messy process by which new species come into existence.

7

The Code(s) by Which You Live

AS WE HAVE SEEN, ALL LIVING THINGS are related, in the sense that they have an ancestor in common. Because it would have lived billions of years ago, you might think that biologists wouldn't have much to say about this ancestor, but you would be mistaken. It was, they tell us, a carbon-based life form and a single-celled organism. They are also convinced that it used DNA to store and transmit its genetic makeup. And significantly, they can tell us the code it used to "translate" genes into proteins. It is this code that we will explore in the pages that follow, but before we do, a bit of review might be useful.

Proteins are best known as the building blocks of your body. Your muscles contain protein, as do the membranes of your cells. Your ligaments, tendons, bones, and skin contain the protein collagen. Your hair is made of the protein keratin, as are your fingernails. The lenses of your eyes are made of crystallin, a water-soluble protein.

Besides playing the role of molecular building block, though, protein molecules can do things. Consider, for example, the muscle protein myosin. You walk by sending a signal to the myosin molecules in your leg muscles. These molecules have little oar-like extensions that attach to filaments of actin, a structural protein. In response to the signal, these little oars pull back simultaneously, thereby causing your muscles to contract. And besides making it possible for you to walk, protein molecules can *themselves* walk, on little molecular "legs." In particular, molecules of the protein kinesin take step after step along microtubules—made of tubulin, another structural protein—towing material from one part of a cell to another. And more amazing still are the proteins that join together to form the rotary motors that allow microbial flagella to spin.

Another thing proteins can do is play the role of molecular chemist. The protein tyrosinase, for example, causes your skin, when exposed to sunlight, to produce melanin, thereby allowing light-skinned people to tan. The protein pepsin, along with other digestive enzymes, tears the protein molecules in the foods we eat into their constituent amino acids. Other proteins, rather than tearing apart molecules, combine them to make bigger molecules. To do this, they bring two molecules together and cause them to touch at just the right place for them to bind.[1]

Consider all the things you can do. You can move? It is because your motor-protein molecules can move. You can see? It is because the opsin proteins in your retina can detect light. You can hear? It is because the protein prestin in your inner ear responds to sound. You can taste and smell the food you eat? It is because your mouth and sinuses have proteins that react to chemicals in foods and in the air. You can fight germs? It is because you have proteins, known as antibodies, that can identify and neutralize pathogens. Remove the proteins from your body, and you would become a pathetically helpless organism—one, I should add, that would soon perish.

Despite their variety and their incredible diversity, your proteins all have the same chemical structure: they consist of amino acid molecules, strung together like the links of a chain. Once a one-dimensional chain has been formed, it folds[2] into a very precise three-dimensional structure that generally looks, at the atomic level, like a tangled chain. The order of the amino acids in a protein is very important since it determines how the chain folds up, which in turn determines how the protein functions. Misfolded proteins are thought to be behind numerous prion diseases, including bovine spongiform encephalopathy, also known as *mad cow disease*, and kuru. Fatal familial insomnia is another prion disease. A misfolded protein prevents its victims from falling asleep. As a result, they experience delusions, dementia, and eventually death.

Your body is capable of making thousands of different proteins, but to make them, it requires only 20 different amino acids as raw materials. That proteins would be capable of doing all the things I have described is incredible, and that your body can make proteins by stringing together only 20 different kinds of molecules is more incredible still. It is hard to study proteins without concluding, as biochemist Nick Lane does, that they are "the crowning glory of life."[3]

TO BE ABLE TO CONSTRUCT PROTEINS, your body needs "recipes" from which to work. The recipes in question are stored in the genes of your DNA. They are "written" not in English or even in chemical symbols, but in chemicals known as *nucleotides*. There are about 20,000 of these gene recipes, but because a single recipe can contain variants, your body is capable of manufacturing far more than 20,000 proteins.[4]

To forestall confusion, it is important to keep in mind that nucleotides are chemically different from amino acids. Consequently, nucleotides are not components of any protein; they are instead components of a coded recipe for a protein. In much the same way,

Amino Acid	DNA codons that represent it	Amino Acid	DNA codons that represent it
Alanine	GCT, GCC, GCA, GCG	Leucine	CTT, CTC, CTA, CTG, TTA, TTG
Arginine	CGT, CGC, CGA, CGG, AGA, AGG	Lysine	AAA, AAG
Asparagine	AAT, AAC	Methionine	ATG
Aspartic acid	GAT, GAC	Phenylalanine	TTT, TTC
Cysteine	TGT, TGC	Proline	CCT, CCC, CCA, CCG
Glutamic acid	GAA, GAG	Serine	TCT, TCC, TCA, TCG, AGT, AGC
Glutamine	CAA, CAG	Threonine	ACT, ACC, ACA, ACG
Glycine	GGT, GGC, GGA, GGG	Tryptophan	TGG
Histidine	CAT, CAC	Tyrosine	TAT, TAC
Isoleucine	ATT, ATC, ATA	Valine	GTT, GTC, GTA, GTG

FIGURE 7.1. This chart shows which DNA codons, consisting of three nucleotides, "stand for" which amino acids. Your body's protein-making machine uses this chart to decipher the nucleotides it finds in protein-coding genes. Notice that the chart is redundant, in the sense that most amino acids are represented by multiple DNA codons. Notice, too, that three codons—TAA, TAG, and TGA—are missing from the above chart. They are "stop codons" that do not stand for amino acids but instead tell your body's protein-making machine that the protein it is constructing is complete.

the word *butter*, although it will appear in a recipe for puff pastry, will not itself be a component of a puff pastry made in accordance with that recipe. The puff pastry will instead contain the yellow dairy product for which the word *butter* stands.

Your body uses four nucleotides to write its protein recipes. They are adenine, thymine, cytosine, and guanine, which generally are identified by their first letter: *A, T, C,* and *G*. Consequently, the recipe for a protein might start out like this: ATGACAACGCTT. To simplify our discussion, we can imagine that your body has little machines that make proteins in accordance with DNA recipes.[5] When one of these machines is given a recipe in the form of a string of nucleotides, it first groups it into segments that are three nucleotides long: ATG-ACA-ACG-CTT. Biologists refer to these segments as *codons*. The protein-making machine then looks these codons up, so to speak, on a biological code sheet[6] to see what amino acids they stand for (see Figure 7.1). In your body's code, ATG stands for the amino acid methionine, ACA and ACG both stand for threonine, and CTT stands for leucine. The protein-making machine then strings together the indicated amino acids in the specified order to make a chain of them: methionine + threonine + threonine + leucine. This chain, when finished, will constitute a protein. See Figure 7.2 for a cartoon that depicts the operation of your body's "protein-making machine." Readers should be warned, though, that the actual procedure by which your body constructs proteins is vastly more complex than this cartoon makes it look.[7]

The cartoon in Figure 7.2 reveals something surprising about the code your body uses to decipher recipes written in nucleotides: it is *redundant*, in the sense that two different codons, ACA and ACG, are used to code for the same amino acid, namely, threonine. They are, in other words, synonyms. And this is only the tip of the redundancy iceberg: ACT and ACC also code for threonine. Furthermore, there are not four, but

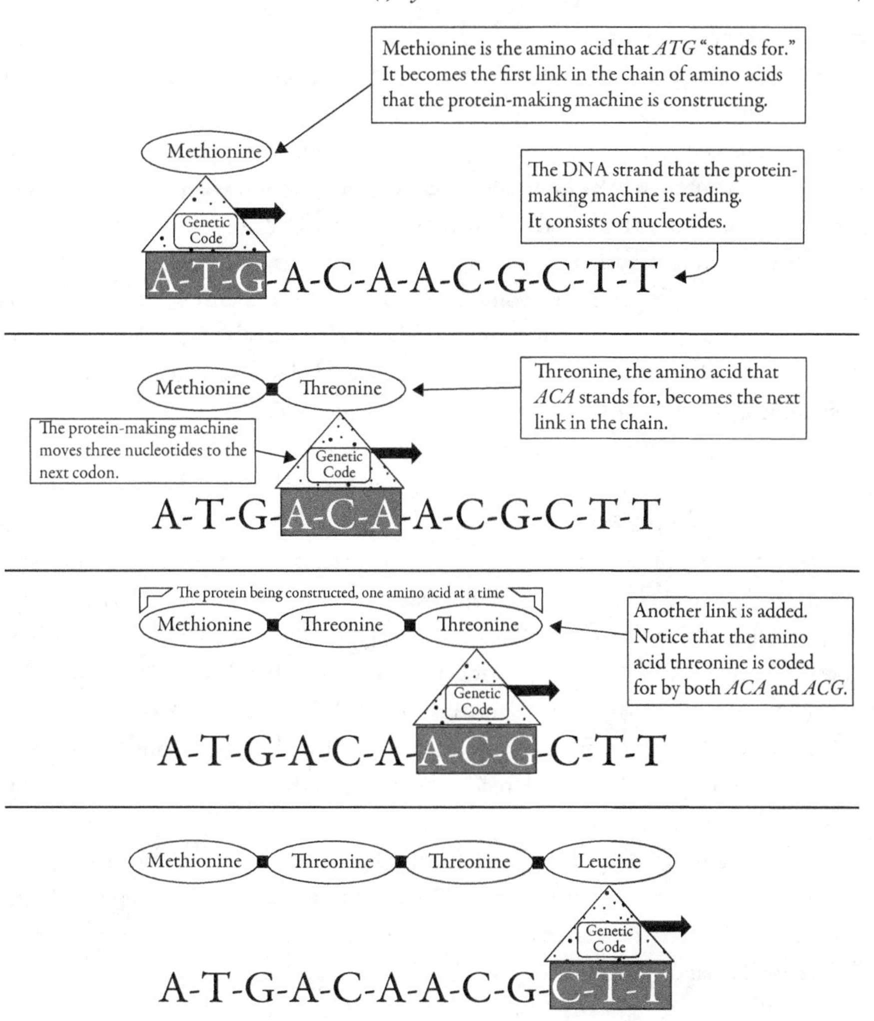

FIGURE 7.2. The "protein-making machine," represented by a triangle, moves along a strand of DNA. It looks up each three-nucleotide-long "codon" on the genetic code—see Figure 7.1—and adds the indicated amino acid onto the chain that it is forming. The resulting chain will constitute a protein, and once it is formed, will fold itself into the three-dimensional shape that determines its function. I should add that the biological process by which proteins are manufactured is vastly more complicated than this and that "the" protein-making machine in fact consists of multiple machines that reside both inside and outside of the cell nucleus.

six different ways to code for the amino acid leucine: CTT, CTC, CTA, CTG, TTA, and TTG.

THE EXISTENCE OF THESE SYNONYMOUS CODONS IS INELEGANT. To see why I say this, consider the codes used by computer programmers. Computer circuits can process only numbers—more precisely, binary numbers. Therefore, if a programmer wants a computer to do something with letters, he has to represent them with numbers. Programmers have therefore developed the ASCII code, in which numbers "stand for" letters. In this code, *65* stands for the letter *A*, *66* stands for *B*, and so on. Suppose, however, that when this code was being devised, a programmer had suggested that *A* be represented not merely by *65* but by *66* as well, and that *B* be represented by, say, the numbers *67*, *68*, *71*, and *88*. His fellow programmers would have derided this code as being foolishly wasteful, a violation of coding aesthetics. On seeing the genetic code shown in Figure 7.1, these programmers would likely respond in similar fashion: "Leucine has six different biological codes? How dumb is that!" Nature, of course, cares not a whit about the opinions of programmers; indeed, it has decided to make the very lives of these programmers depend on a genetic code that they would ridicule.

Besides being inelegant, the redundancy of the genetic code is inconvenient for biologists, inasmuch as it makes it impossible for them to deduce the DNA recipe for a protein from the order in which amino acids appear on that protein. To see why, consider again the very simple protein shown in Figure 7.2. It is only four amino acids long. Thanks to the redundancy of the genetic code, though, there are 96 different recipes which, if followed, would yield this same protein. As we have seen, the recipe used to construct it was ATG-ACA-ACG-CTT. It could just as well have been ATG-ACG-ACA-CTT (since ACA and ACG are synonyms), ATG-ACA-ACG-TTA (since CTT and TTA are synonyms), or 93 other recipes.[8] Consequently, biologists cannot "work backward" from proteins to deduce the recipe that was used to construct them. They must instead read the DNA recipe itself.

In recent years, biologists have gained sufficient understanding of the genetic code to be able to manipulate it with various objectives in mind. In one experiment, they altered the DNA of *E. coli* bacteria in a manner that would "free up" the codon CCC.[9] They knew that since the codons CCC and CCG both code for the amino acid proline, replacing all of an *E. coli* cell's CCC codons with CCG codons would not affect the proteins it produced and therefore would not affect the cell's functions. In another experiment,[10] biologists in effect added two letters to the genetic code's alphabet, thereby dramatically expanding DNA's ability to store data.[11] In the next chapter, we will consider yet other research that has been done by these genetic engineers.

Genetic mutations alter the nucleotide letters in DNA, which in turn can alter the proteins that an organism produces. Suppose that a protein-coding gene contains ACT, the codon that stands for the amino acid threonine. Suppose that as the result of a mutation, one letter of this codon changes. If it is the *last* letter that changes, the ACT codon will transform into ACA, ACC, or ACG. Because these codons also stand for threonine, this mutation will have no impact on the protein that is produced: at that point in the chain that is being constructed, the amino acid threonine will appear. If the *first* letter changes, though, the ACT codon will transform into CCT, GCT, or TCT, and a completely different amino acid will be added to the chain: CCT codes for the amino acid proline, GCT for alanine, and TCT for serine. Replacing threonine with proline, alanine, or serine will result in an entirely different protein that will very likely function differently than the original protein did. This in turn can have a significant impact on the organism that experienced the mutation.

Along these lines, consider normally dark-haired Melanesians. A change in a single letter in the section of DNA where the gene recipe for one of their hair proteins is stored will change one of the amino acids that appears in that protein, which in turn will result in them being blond headed.[12] A rather different single-letter change in DNA played a profound role in human history. The wild grass known as *teosinte* is an ancestor of modern corn. There is evidence that at some point in the past, the letter *C* took the place of the letter *G* at one specific place in teosinte's DNA.[13] As a result, the codon AAG, which stands for the amino acid lysine, was transformed into AAC, which stands for aspargine. This slight genetic change had the effect of softening the thick, hard case of teosinte kernels. Modern corn has inherited this mutation, and as a result, its kernels are much more digestible than those of its teosinte ancestors. To fully appreciate the impact of this single-letter change, imagine a world without modern corn as a foodstuff.

A BIT OF MATH REVEALS THAT if we have four nucleotides to work with and if codons are three nucleotides long, 64 different codons are possible.[14] More math reveals that there are 1.5×10^{84} different ways that 64 codons can pair up with the 20 amino acids that your body uses to construct proteins, which in turn means that 1.5×10^{84} different genetic codes are possible.[15] This is more genetic codes than there are atoms in the universe. Nevertheless, your body and every other currently living thing uses the same genetic code. Why should this be?

It is not because this is the only code that works; they all do. There is no reason, for example, why ATG has to code for methionine and ACA has to code for threonine; these codes could have been interchanged, and biological processes would carry on as usual, as long as all the ATG and ACA codons in your DNA were also interchanged. In similar fashion, the ASCII code used by programmers could have switched the codes for *A* and *B*, with *65* standing for *B* and *66* standing for *A*, and computers could still process words. The conclusion we are led to is that the genetic code, besides being redundant, is incredibly arbitrary.

Besides inheriting DNA from their ancestors, organisms inherit the code with which to read that DNA. This means that there are two ways in which two organisms could come to use the same genetic code. One is if they have an ancestor in common from which they inherited that code. The other is if, although they have no ancestors in common, they inherited their genetic code from two ancestors who, in the process of evolving, coincidentally hit on the same code. Given the arbitrariness of the genetic code, though, such a coincidence is extremely unlikely—indeed, it is even less likely than a fair coin coming up heads 279 tosses in a row.[16] This is why the arbitrariness of the genetic code, taken together with its universality, counts as "smoking-gun" evidence that all currently living things have an ancestor in common.[17]

To better understand the logic of this argument, consider the Morse code, which was used back in the days of telegraphs. This code, invented by Samuel Morse, consisted of dots and dashes, grouped into clusters and separated by pauses. To send an SOS signal, a telegraph operator would send three dots, for the letter *S*, followed by a pause, then three dashes for the letter *O*, followed by a pause, followed by three more dots. To send the word *COW*, the operator would send dash-dot-dash-dot for *C* followed by three dashes for *O*, followed by dot-dash-dash for *W*. One might think that dot would be the obvious symbol for a period, the mark of punctuation that indicates the end of a sentence, but it wasn't. Instead, the code for a period was dot-dash-dot-dash-dot-dash, with the single dot being reserved for *E*, the most-used letter in the English language.

Realize that the Morse code is quite arbitrary: it would still work if the codes for *O* and *S* were reversed, with dot-dot-dot standing for *O* and dash-dash-dash standing for *S*, as long as everyone using the code knew of the switch. Indeed, the code for every letter could be switched, and the code would still work.

Suppose, then, that one of Samuel Morse's contemporaries also claimed to have invented a telegraphic code, and suppose it turned out to be identical to Morse's. Because the assignment of dashes and dots to letters is highly arbitrary, it is unlikely that two inventors would have come up with the same code independently. Therefore, under these circumstances, the logical conclusion to draw is that one of the inventors had copied the other's code. In much the same way, the arbitrariness of the genetic code, taken together with its universality, is compelling evidence that all living things have copied the genetic code of a common ancestor, meaning that they are all related.

We will turn our attention to this ancestor in the next chapter. Before doing so, however, let us pause to deal with a wrinkle in this story. It turns out that some organisms use slight variants of the code given in Figure 7.1. Indeed, your body itself makes use of two different genetic codes. The DNA in the nuclei of your cells is deciphered using the coding scheme shown in Figure 7.1, but the DNA in your cells' mitochondria—I will have much more to say about these little organelles in chapter 11—is deciphered using a slightly different genetic code in which, for example, the codon ATA, instead of coding for the amino acid isoleucine, codes for methionine. In other words, you live in accordance with not one, but two (slightly) different genetic codes. And this is only one of many exceptions to the "universal" genetic code.[18] The exceptions in question, however, are minor, and they are what we would expect from an evolutionary process that is perfectly willing to tinker with its designs for living. These exceptions do not, therefore, undermine the argument, based on the genetic code(s), that all currently living things have an ancestor in common.

8

Your (Alien?) Roots

IN CHAPTER 4, WE EXAMINED THE TREE OF LIFE. We saw that all the branches of this tree converge on an organism that biologists have labeled *LUCA*, the Last Universal Common Ancestor of all currently living things. It is tempting to think that LUCA also would have been the first living thing on Earth, but it couldn't have been. The genetic code it used—and that we and every other currently living thing have inherited—is too complex to have popped into existence out of nowhere. Instead, LUCA would have inherited its genetic code from ancestors that evolved it over many generations, by means of the trial-and-error process used by evolution. Going back in time, these ancestors would presumably have had ever-simpler genetic codes.

In the tree of life presented in Figure 4.1, LUCA appears at the top of the "trunk" of the tree, and all of our planet's currently existing species can trace their origin back to it. At the bottom of this trunk is the "first living organism,"[1] from which LUCA evolved. Below the trunk are "roots" to indicate the semi-animate stuff from which this organism evolved. This depiction of the tree of life is misleading, though, inasmuch as it makes it look like there was "straight line evolution" from the first living organism to LUCA, when in fact, there would have been ongoing evolutionary experiments during that period, and as a result, the trunk would have had many branches emerging from it. Figure 8.1 shows such a tree.

Back in chapter 2, we saw that MRCA, the Most Recent Common Ancestor of all currently living people, probably would not have been the only person alive at the time he or she—or they—existed. It is similarly likely that LUCA existed contemporaneously with other species. Had these "cousins" of LUCA not gone extinct, though, they would have

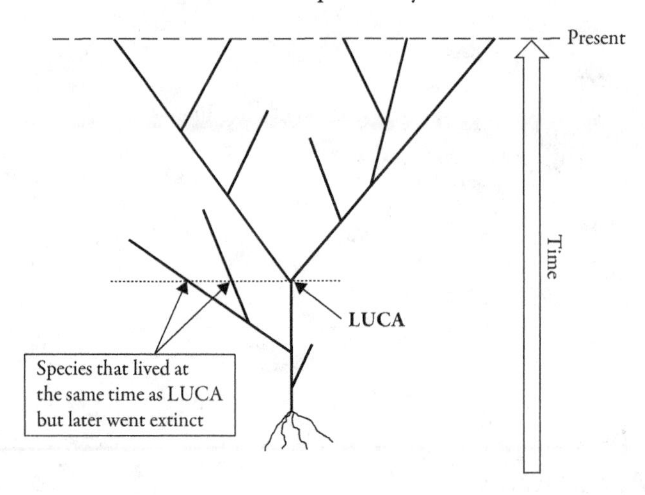

FIGURE 8.1. The tree of life, revisited. LUCA, the Last Universal Common Ancestor of all currently existing species, would probably have had "cousin species" that branched off the "trunk" of the tree before LUCA appeared. These cousins would have had ancestors in common with LUCA but because they went extinct would have no present-day descendants.

deprived the organism we are calling *LUCA* of the LUCA title. That title would instead go to the ancestral organism that LUCA and its cousins had in common. Realize, too, that had we been around at the same time as the organism we call LUCA, we could not have known that it would gain the LUCA title, since this would be determined by the future reproductive activities of the earth's then-living organisms.

Significantly, it is possible—likely, even—that the life experiment that resulted in the existence of LUCA is only one of many such experiments that took place on early Earth. It is possible, in other words, that life has arisen many times on Earth, only to subsequently perish. If this is the case, though, a correct representation of life on Earth will show not one tree but multiple trees of life (see Figure 8.2). The organisms in these other trees might not have used DNA to store their genetic information,[2] and if they did, their genetic codes would likely have been radically different from our own. These alternate trees of life perished, however, when the species represented by their branches all became extinct.

In making this last statement, though, I am being presumptuous. It is possible that the rival trees of life did not *all* die. It is possible, in particular, that some of the organisms now alive can trace their ancestry back to a life experiment other than the one that gave rise to us. It is also possible that if such "alien" organisms did exist, biologists would be oblivious to them. They might look very much like "normal" microorganisms, and it would only be if a biologist isolated one of them in order to analyze its DNA that she would discover, much to her astonishment, that it used a dramatically different genetic code than we do,[3] or even more shocking, that it didn't have any DNA to analyze!

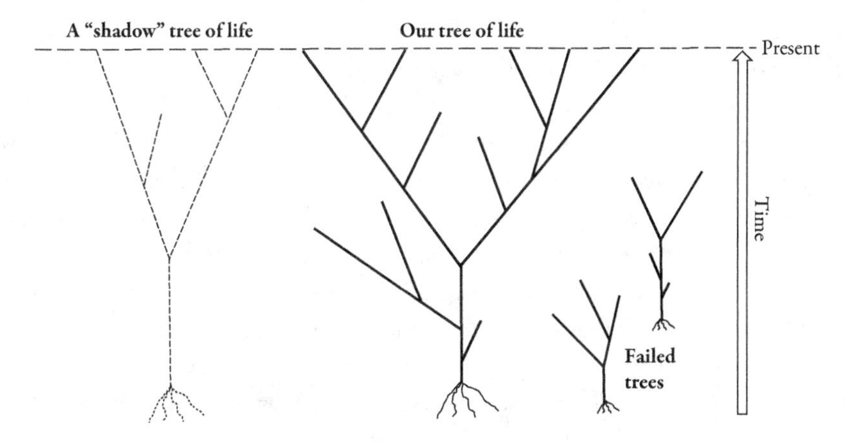

FIGURE 8.2. It is possible that life has arisen many times on Earth. If this were the case, then besides our tree of life, there will be others. We will have no ancestors in common with the species on these alternate trees. If the species on an alternate tree have all gone extinct, it will be a failed tree of life, but if some of its species still exist but have not yet been discovered and revealed for what they are, it will be a "shadow" tree of life.

If alien organisms do exist, then an accurate tree-of-life diagram, besides showing our tree of life and various failed trees of life, will show one or more "shadow" trees.[4] Because these trees are the result of different life experiments, they will have different "root systems" than our tree does. Furthermore, because these trees represent life experiments that are still in progress, their branches will extend to the present time, the way those of our tree do. Figure 8.2 shows how a shadow tree of life, if one exists, would be depicted.

LET US NOW TURN OUR ATTENTION to the trunk of the tree of life. LUCA is at the top of this trunk. Below it are LUCA's ancestors, and the further down on the tree of life we go, the simpler their genomes become. Your genome consists of DNA that is, in total, 3.2 billion base pairs long.[5] A genome this long and complex, though, could not simply pop into existence. It is instead the result of additions of DNA to a shorter genome, which itself will be the result of additions of DNA to an even shorter genome.

To gain insight into how short the genome of a living thing can be, it is useful to consider the genomes of currently living organisms. The shortest known genome is that of *Nasuia deltocephalinicola*, a bacterium with only 112,000 base pairs and 137 protein-coding genes (compared to humans' 20,000 protein-coding genes).[6] But although it is alive, *N. deltocephalinicola* is not "free-living." Because its ancestors have spent so many generations parasitizing leafhoppers, *N. deltocephalinicola* bacteria have lost the ability to produce many of the substances they need to live, preferring instead to take them from their host. Indeed, remove them from a leafhopper and they quickly perish. The smallest known genome in a free-living organism—one that can live outside a host—is that of *Mycoplasma genitalium*, which has only 517 protein-coding genes.

To gain further insight into how short the genome of a living thing can be, biotechnologist Craig Venter and his colleagues have created "synthetic" microbes. Their experiments exploit the free-living bacterium *Mycoplasma capricolum*. Its genome, which includes 901 protein-coding genes, is carried on a single circular "bracelet" of DNA known as a *bacterial chromosome*. If biologists remove this chromosome, the *M. capricolum* cell is deprived of its "recipe book" for making proteins and as a result will perish, but they have found that if they replace it with the chromosome of a *Mycoplasma mycoides* bacterium, the *M. capricolum* cell will start making the proteins in the *M. mycoides* recipe book and will thereby continue to live. They christened the resulting organism, with the body of one bacterium and the DNA of a bacterium of a different species, *Syn 1.0*, a reference to its synthetic nature. These same scientists then started systematically removing genes from the chromosome of the Syn 1.0 organism. When they got it down to 516 genes—one fewer than *M. genitalium* has—they rechristened their organism *Syn 2.0*. They have continued to shrink its genome and in 2016 announced the creation of Syn 3.0, which has only 473 genes.[7]

Two comments are in order before we move on. The first is that although scientists have gained an understanding of what genes do, they are in the dark about the function of one-third of Syn 3.0's genes.[8] This is an indication of how much more we have to learn before we can claim to understand the life process. The second comment is that contrary to some headlines about Venter's work, he has not created life. In particular, in his creation of Syn 1.0, he did not bring inanimate matter to life; he simply altered an organism that was already alive. It is a very important difference. It is clear that biologists have a lot more work to do in order to explain how life could transform, through evolutionary processes, from organisms with tiny genomes into organisms as complex as LUCA presumably was.

THIS BRINGS US TO THE BASE OF THE TRUNK of the tree of life. The organisms we find there would have had very simple DNA—or maybe none at all. They might instead have relied on RNA to carry on the life process. Allow me to explain.

DNA molecules are remarkable for their data-storage ability. Indeed, it has been estimated that if various technical difficulties could be overcome, one gram of DNA could store all the data currently stored by every major tech company, including those data-storage goliaths Google and Facebook, and still have room to spare.[9] Furthermore, under the right conditions, DNA can store that data for thousands of years. Having praised DNA in this manner, though, I need to add that it is a one-trick pony: yes, it does a superlative job of storing data, but this is the *only* thing it can do. In particular, a DNA molecule can't, all by itself, move or initiate chemical reactions, it can't make proteins, and it can't make copies of itself. All it can do is sit there, storing data, much like a compact disk does. It is, to be sure, ironic that the "molecule of life," as some have called it, would itself be singularly lifeless.

As we have seen, DNA contains, among other things, coded recipes for the manufacture of proteins. But your body's protein-making machine, described in chapter 7, is itself composed partly of proteins.[10] Consequently, unless proteins exist, DNA's protein recipes are unreadable; but unless the protein recipes exist, your body won't know what proteins to make and in particular won't know how to make the proteins used by the protein-making machine. So which came first, the proteins used by your body's protein-making machine, or the DNA recipes for the assembly of those proteins?

And this isn't the only paradox involving DNA. The protein recipes in your DNA are, as we have seen, written in code. Consequently, in order to make proteins in accordance with these recipes, the protein-making machine must first decode them. To do so, it consults the genetic code that is stored—where else?—in your body's DNA, in the form of what are known as *tRNA genes*. In other words, the genetic code necessary to decode the genetic code is itself coded.

To better appreciate this paradox, suppose someone handed you a coded message with a request that you decode it. Your response: "Fine, but I'll need to have the key for the code." Their response: "You already have the key, there in your hand. It is in the first part of the coded message." Such a key obviously wouldn't be very useful, since you could gain access to it only if you already knew the code, in which case, you wouldn't need the key. How very strange—and irksome.

Some creationists take the existence of paradoxes like these as evidence that life could not evolve without divine intervention.[11] Only an all-powerful, all-knowing being, they claim, could successfully resolve what would otherwise be insoluble chicken-and-egg situations. Before joining creationists in taking such a drastic measure, though, we would do well to consider the history of computers. Modern computers are so complex that it takes computers to design them. But if it takes a computer to design a computer, how could computers have arisen in the first place?

Since we know the history of computers, we have a ready solution to the chicken-and-egg problem raised by this question. Early in their "evolution," it *was* possible to design a computer without resorting to computers. As computers became more complicated, though, it became possible and convenient for engineers to use computers to design computers, and not long after that, computers were sufficiently complex that their design was impossible without the use of computers. Something very much like this probably happened early in the history of life, and RNA probably played a central role in this early history.

COMPARED TO DNA MOLECULES, RNA molecules are spectacularly talented. In chapter 7, I described your body's protein-making machine in much-simplified terms, saying that it manufactures proteins on the basis of recipes found in your DNA. This machine in fact relies heavily on RNA. The first step in the protein-manufacturing process is for a molecule of RNA polymerase to copy a gene onto a data-carrying "note" consisting of messenger RNA.[12] Elsewhere in the cell, an RNA-containing ribosome deciphers the note. The "key" it uses to do so is contained not in a chart like the one shown in Figure 7.1 but in molecules of transfer RNA. The ribosome then strings amino acids together, in accordance with the deciphered note, to make a protein. And to this impressive list of talents, we need to add one more: RNA molecules are capable of making copies of themselves without outside assistance.[13]

The versatility of RNA molecules has led biologists to conclude that RNA-based life forms preceded those that made use of DNA.[14] We would therefore find these RNA-based life forms at the bottom of the trunk of our tree of life. With the passage of time, these organisms presumably started using DNA for data storage, and with the passage of even more time, they would have evolved into organisms dependent on both DNA and RNA, the way we are. This *RNA World hypothesis*, as it is known, takes us back an important step in our attempt to discover the roots of the tree of life but still leaves us with the question of how these RNA-based life forms, which themselves would have been complex, arose.

Biologists have proposed many theories in response to this last question. They agree on the importance of water to the formation of life,[15] but they disagree about where that life first formed. They unanimously reject Charles Darwin's suggestion that life arose in a warm little pond and instead speculate on the possibility that it arose in thermal vents on the ocean floor—which is where, as we have seen, LUCA probably lived—or maybe in clay or on the surface of a crystal. They also disagree about how the first living things would have made a living—about, in other words, the metabolic process by which they extracted energy from their environment.[16]

However it happened, the evolution of life would have been a gradual process. In the same way as there isn't a point in time at which one species transforms into another, there won't be a point in time at which we could declare that life had arisen. If we visited Earth 4.3 billion years ago, we would probably be confident that there were no living things. If we visited half a billion years later, we would probably be confident that very simple life forms existed. Between those two times, however, we would find it difficult to declare either that life existed or that it didn't. Instead, we might characterize the things we saw around us as being "lifelike."

One thing is clear, though. About 4 billion years ago, an organism found itself in possession of the "spark of life." In the same way as an ember from one fire can be used to start another, that organism passed this spark to her daughter cells, which in turn passed it on to theirs, and so on, down to this very day. Indeed, while you read this sentence, new cells are coming into existence within you. These cells will inherit not only some of the matter from the cell that fissioned to make them, along with its genome, but the spark of life as well. In other words, you carry within your body the flames of a fire that has burned, in both the metaphorical and metabolic sense of the word, for 4 billion years. You are, to be sure, a living organism, but it is probably more accurate to describe you as a current manifestation of the life force.

SO FAR IN OUR DISCUSSION OF THE TRUNK of the tree of life, we have assumed that this tree will have its root here on Earth. Some have suggested, though, that to find the source of life on Earth, we should turn our attention not to, say, thermal vents on the ocean floor, but to distant planets. According to this *panspermia theory*, as it is known, life arose elsewhere in the universe and, having arisen, was somehow transported here. There are different ways this might have happened. Perhaps an asteroid struck another planet and blasted pieces of it into space, and those pieces and the living organisms they carried ultimately ended up on Earth. Or maybe alien beings—ancient astronauts, if you will[17]—visited Earth in the past and brought life with them.

If the latter were the case, the life in question could have been brought to Earth accidentally, the way a lab assistant might accidentally contaminate a petri dish. All it takes for life to blossom is for one microorganism to find itself in an agreeable environment. Alternatively, the contamination might have been intentional. Aliens might, for example, have inoculated the earth with living organisms as part of a biological experiment. And taking this line of thought one step further, there is an outside chance that life on Earth is the result of an over-achieving alien kid's science fair project.

Panspermia theory, to be sure, would be difficult to prove or disprove. One compelling piece of evidence for its truth would be if aliens came to Earth and announced that they were responsible for life on our planet. If we were skeptical of their claims, we could check their story by asking them for a tissue sample. Suppose our analysis revealed that they were a DNA-based life form. This alone wouldn't be conclusive proof that we were related: perhaps use of DNA for storage of genetic data evolved independently elsewhere in the universe. Suppose, however, that their cells used the same genetic code as ours use to construct proteins on the basis of DNA recipes—or some slight variation of our code. In chapter 7, we saw that such an event would count as smoking-gun evidence that two currently existing terrestrial organisms have an ancestor in common. In much the same way, it would count as smoking-gun evidence that we and our alien visitors have an ancestor in common, thereby lending credence to their claim that they are responsible for our existence.

Along similar lines, suppose that instead of being visited by aliens, we find life on Mars. Suppose these Martian organisms, which would most likely be microbes, share our genetic code. This would leave us with an interesting set of possibilities. Either Mars was the source of life on Earth, Earth was the source of life on Mars (perhaps we accidentally contaminated it with one of our Mars landers), or some third cosmic location was the source of life on both Earth and Mars. It would be hard to know which of these possibilities was true, but given the similarity in our genetic codes, one of them would have to be.

Of course, even if panspermia theory is true, it is intellectually unsatisfying, inasmuch as it doesn't get to the bottom of things. It is like announcing that your ancestors came from Sardinia. As we saw in chapter 1, any thoughtful person, on hearing this claim, will want to know where *those* ancestors came from. All panspermia theory does is shift the question of how animate organisms could have arisen from inanimate stuff here on Earth, to the question of how this could have happened on some distant planet.

Let us, for the sake of argument, suppose that Earth's life arose here on Earth. It would have been a remarkable event. Our planet formed 4.56 billion years ago and at that time would have been a molten sphere. Over the next few hundred million years, Earth cooled enough for it to form a crust, this crust cooled enough for it to be habitable, and then, quite remarkably, life somehow arose "from scratch" on that crust.

When we examine life on Earth, we find compelling evidence of its hardiness and how opportunistic it is. There are living things pretty much everywhere we look. They exist not only deep in the ocean but under the seabed and around superheated oceanic geysers. There is life in the air, not only in the form of birds that can fly as high as Mount Everest but in the form of airborne microbes. There is life not only in the earth's soil but in its rocks—indeed, in cracks in the bedrock thousands of feet below its surface. It takes effort to find places on Earth that are devoid of life.

If life can arise on Earth, surely it has done so elsewhere. The universe is, after all, huge. It has hundreds of billions of galaxies. Many of these galaxies in turn have hundreds of billions of stars, each of which probably has multiple planets—and probably moons as well—on which life can potentially arise. Therefore, even if there is a one-in-a-billion chance of life arising on a planet like Earth, there are likely trillions of places where it can arise, making it highly probable that life would arise on many of them. Keep in mind, too, that our universe has been around for 13.8 billion years. That is a lot of time for nature to conduct life experiments, increasing the chance that some of them will be successful.

To better understand the logic of this argument, consider a lottery in which bettors have to pick a series of four numbers between 0 and 9. There will be 10,000 such series to choose from, including 0000, 9999, 4758, 8514, and so on. The chance of 7777 coming up in this lottery is therefore 1 in 10,000, which is pretty remote. But if we repeat this lottery a million times, 7777 is likely to come up on 100 occasions. Similarly, even though it may have been a long shot that life would arise on our planet, it is almost certain that it would have arisen on many of the trillions of planets that exist. Ours, as it so happens, was one of them.

The winners of a pick-four lottery will be amazed by their success and will probably come up with reasons for having beaten the odds. Perhaps some higher force is compensating them for past injustice or rewarding them for being such wonderful people? But the rest of us will see these winners for what they are: the lucky fools who won the lottery. Consistency demands that we should likewise see our biological selves for what we are: the very lucky fools that won the life lottery.

IN THIS CHAPTER, WE HAVE TALKED about the possibility of life on Earth having arisen as the result of a visit by alien beings. Let us end the chapter with a bit of speculation about what such aliens, if they existed, would be like.

In many science fiction movies, earth-visiting aliens are about the same size and shape as humans. In the past, there was a practical reason for this: the beings in question would have been played by humans wearing costumes. This is no longer the case, though. Because cinematic aliens are these days generated by computers, they can be almost any size or shape you can imagine. Some shapes and sizes, however, are more plausible than others. In particular, in order to build an interstellar spacecraft—or even a radio transmitter—aliens will have to be able to skillfully manipulate their environment. This will require them to have appendages and sensory organs. These aliens will also have to be of a certain size: too small, and they won't be able to move things around; too large, and they will be clumsy. As we saw in our discussion of the Einstein whale in chapter 3, if you have the wrong sort of body, there will be all sorts of things you can't do, even if the brain within that body is incredibly intelligent.

At this point, a science fiction fan might suggest that aliens don't need to build spacecraft; they can instead have their robots build them. This is an intriguing suggestion, but it immediately raises an obvious question: who built their robots? Aliens who have the wrong sort of body for building spacecraft will also, one supposes, have the wrong sort of body for building robots. This is something to keep in mind the next time you watch a movie about the earth being invaded by alien beings. Do those aliens, regardless of how intelligent they are, have bodies that would allow them to build the spacecraft that brought them to Earth, or to build robots that could build those spacecraft?

Besides drawing conclusions about the bodies of alien beings capable of traveling to Earth, we can draw conclusions about their minds. They would obviously have to be very intelligent, but I would argue that any alien being intelligent enough to build a spacecraft capable of an interstellar voyage would be intelligent enough to know that it would be a waste of their time and energy to make such a voyage. To see why I say this, consider Proxima Centauri, the star that is the sun's closest neighbor. Suppose it has a planet that harbors intelligent life, and suppose its inhabitants decide to come visit us. If they travel in rocket ships, the way we do, it might take them a hundred thousand years to reach Earth.[18] That would be a long time to spend aboard a spacecraft. Would any sensible being agree to go on such a trip? And if they did, would they or their descendants be able to survive it?

To put this journey into context, recall that we earthlings have traveled to the moon. The trips required only three days of travel in each direction. Nevertheless, after making six such trips, we decided that, all things considered, we had better things to do with our resources. And realize that whereas the moon is 1.3 light-*seconds* away, Proxima Centauri is 4.3 light-*years* distant.

Rather than come here in person, it would be much more sensible for Proxima Centaurians, if they exist, to send an unmanned satellite, the way we have in our exploration of the other planets of the solar system. But even doing this would represent a leap of faith on their part. Such a satellite also might take a hundred thousand years to reach Earth. When it did, it would dutifully—if it was still functional—radio back data about what it found. It would take only 4.3 years for these messages to reach home, but would anyone be there to receive them? A hundred thousand years is, after all, a long time for a civilization to exist, and even if it does, will the distant descendants of today's Proxima Centaurians even remember that a satellite was launched?

At this point, ardent fans of science fiction, who want in the worst way to come into physical contact with alien beings, typically start talking about the worm holes and hyper-drives that they believe will enable aliens to cross galaxies in about the same amount of time as our cars cross states. It makes for a nice story, but I'll believe in such things when I see them.

In conclusion, alien life, even intelligent alien life, doubtless exists, but because of its intelligence, it can think of better things to do with its time and resources than visit Earth. Remember this the next time someone tries to tell you about the ancient astronauts who came here to teach the Egyptians how to build pyramids or who, while we sleep, are out in our fields constructing intricate crop circles.

PART II

The Cellular You

9

You Are Complex

YOU BEGAN LIFE AS THE FERTILIZED EGG CELL known as a *zygote*. It came into existence as the result of a rather complex process. In most cases, your mother and father engaged in sexual intercourse.[1] Your father's sperm started swimming up your mother's reproductive tract. Half of these sperm would have been carrying an X chromosome, and the other half would have been carrying a Y. Whether you ended up a male or female depended on whether it was an X- or Y-carrying sperm that made it to the egg.

At one point in their journey, these sperm had an important decision to make: take the left fallopian tube or the right? Since only one of these tubes would, under normal circumstances, lead to a ripe egg, half of the sperm would have chosen the wrong tube, meaning that the journey they were about to take was in vain. Even those that chose correctly, however, would have encountered many obstacles in their subsequent travels. They might, for example, have gotten trapped in the villi that line the fallopian tubes. They would also be swimming against the currents created by those villi. On making it to the egg, they would first have to make their way through its corona and then its cell wall. Although your mother may have said "Yes" to your father, her reproductive system seems to have been saying "No way!" It made it hard for sperm to reach the egg, presumably to ensure that those that did would be healthy. The sperm that ultimately penetrated the egg would have won a race against a few hundred million other sperm.

Realize that the outcome of this race was incredibly random. Had the sex act been carried out in a slightly different way or at a slightly different time, the resulting baby would have had different genes and might have been born a different sex than it was, meaning that *you* would not have been born. Of course, if someone other than you had been born, you wouldn't now exist and that person would be the "you" I would be addressing. That person would also be the one thanking his or her lucky stars for having come into existence.

AFTER YOUR FATHER'S SPERM CELL had merged with your mother's egg cell, another merger took place: the nucleus of the sperm cell merged with the nucleus of the egg cell.[2] It was in this second merger, which might have happened hours after the first, that the fertilized egg cell acquired a new genetic identity, different from that of your mother, your father, and every other living person—unless, of course, you are an identical twin.

Once this second merger was complete, your zygote started dividing. First its DNA replicated, then two nuclei formed to contain the two copies of it, and finally the cell pinched itself in two, with the two nuclei ending up in the two resulting cells. One cell thereby became two. When a cell divides in this manner, it is tempting to think that a mother cell is giving birth to a baby cell, but it would be a mistake to give in to this temptation, since any subsequent labeling of the cells as mother and baby would be arbitrary: there is no biochemical way to distinguish between the mother and the baby. They are therefore, as we have seen, referred to as *daughter cells.*

Over the next week or so, your cells kept dividing as they made their way down the fallopian tube toward the uterus. During this time, they received no nourishment from your mother. By the time the resulting cluster of cells, now referred to as a *blastocyst*, attached itself to the uterine wall, it consisted of perhaps a hundred cells. But how, you might wonder, can one cell become a hundred in the absence of nourishment? According to the Law of Conservation of Mass, matter can't simply come into existence. Wouldn't the transformation of the single-celled zygote into a hundred-cell blastocyst violate this law?

No, since the cells that divide become progressively smaller. If you cut a pound of cheese into a hundred cubes, you have the same amount of cheese as you started with, but each cube is much smaller than the original chunk was. Because the initial egg is, in cellular terms, really big—it is visible to the naked eye—it can "afford" to divide many times. To better understand this point, consider the ostrich egg. It is a single cell that might weigh three pounds, and yet out of it comes a baby ostrich, the result of many fissions of that cell. And of course, each cell in this baby ostrich is much smaller than the original egg.

On reaching the uterus, you had your first meal. More precisely, your cells took up the nutrients that your mother passed to them through the uterine wall. She in turn got these nutrients from the food she ate. She adjusted to this new demand for her bodily resources by "eating for two."

THERE ARE, TO BE SURE, VARIATIONS on this theme. Sometimes a woman has not one but two or more eggs in play at the same time, and they might come from the same or different ovaries. The eggs in question might then be fertilized by two different sperm. The resulting zygotes will not be identical; they will instead be fraternal twins. They can be of the same sex or opposite sex, and they will be as similar as non-twin full-siblings are—unless, of course, they have different fathers, in which case they will be as similar as any two half-siblings are.

It is also possible for a blastocyst to divide on the way to the uterus. In this case, identical twins will result. They will have the same DNA and be of the same sex—usually.[3] If the halves of a divided blastocyst remain in contact after it divides, conjoined identical twins will result. And finally, it is possible that, after a blastocyst has broken in two, one of the halves will again break in two, resulting in identical triplets.

Besides one embryo becoming two, two embryos can merge to produce what is known as a *chimera*, an organism, the cells of which are genetically mixed. This is how a 70-year old father of four came to have a womb.[4] Doctors discovered it while operating on his hernia. He had apparently shared his mother's womb with a sister whose embryo was subsequently engulfed by his. Her womb thereby became his womb.

And even stranger things can happen. In 2002, Lydia Fairchild was the mother of two with another child on the way. A breakup with Jamie Townsend, her partner, resulted in her applying for welfare. As part of the application, Fairchild, Townsend, and their children had to undergo genetic testing to establish parenthood. The results showed that Townsend was the children's father but that Fairchild was not their mother. Welfare officials were therefore understandably skeptical about her maternity claims. Consequently, when she had the baby she was carrying, a court officer was present to do genetic testing. Again, the tests indicated that Fairchild was not the baby's genetic mother. Authorities concluded that she must be acting as surrogate mother—that the egg of another woman, after being fertilized by Townsend's sperm, had been implanted in her.

Doctors finally realized that Fairchild was a chimera. She had been one of two fraternal twins in her mother's womb. The other twin, who was also female, had not only been engulfed by Fairchild's embryo, but its ovaries had lodged themselves where Fairchild's ovaries should be. As a result, the eggs Fairchild was releasing each month were in fact the eggs of her unborn twin. Although Fairchild was the birth mother of her children and in some sense their biological mother as well, it was the unborn twin who was their genetic mother. Strange but true!

Chimerism, it turns out, is not as rare as one might imagine. When a woman is pregnant, it is quite likely that cells from the fetus she is carrying end up in her body. Likewise, some of her cells might end up in the fetus. Consequently, although the mother and her offspring parted company decades before, each might carry within them a tiny bit of the other.[5] This condition is known as *microchimerism*.

YOUR CELLS DIDN'T JUST POP INTO EXISTENCE. Each cell came from a cell, which itself came from a cell, and so on. This means that in the same way as we can construct the family tree that shows who your ancestors are, we can, in theory, pick any of your cells and construct the "family tree" that shows the cells from which it descended. For the sake of clarity in the remarks that follow, I will refer to the former tree as a *personal family tree*, since it shows relationships between persons, and I'll refer to the latter tree as a *cellular family tree*.

Whereas people come into existence as the result of sexual reproduction, in which two people come together to make one new person, cells almost always come into existence as the result of cell division,[6] in which one cell divides into two daughter cells. This means that a cellular family tree will have a different structure than a personal family tree. This last tree, as we have seen, will resemble a circuit diagram, with double the number of entries each level back (see Figure 4.1). A cellular family tree, by way of contrast, will resemble a bamboo stalk. At the bottom will be the cell whose tree it is. Above that cell will be its mother cell, above which will be its mother's mother cell, and so on. See Figure 9.1 for an example of a cellular family tree.

Besides having a personal family tree, you have a personal tree of descendants that shows your offspring (if you have any), the offspring of your offspring, and so on (see the bottom half of Figure 1.1). In similar fashion, each of your cells, besides having a cellular family tree, has a cellular tree of descendants that shows its daughter cells (if it has any), the daughter cells of its daughter cells, and so on. As we have seen, a person in a tree of descendants can have any number of branches below him; a truly prolific man might have a hundred. Each cell in a cellular tree of descendants, though, will have exactly two branches below it, for its two daughter cells, and below each of those branches might be

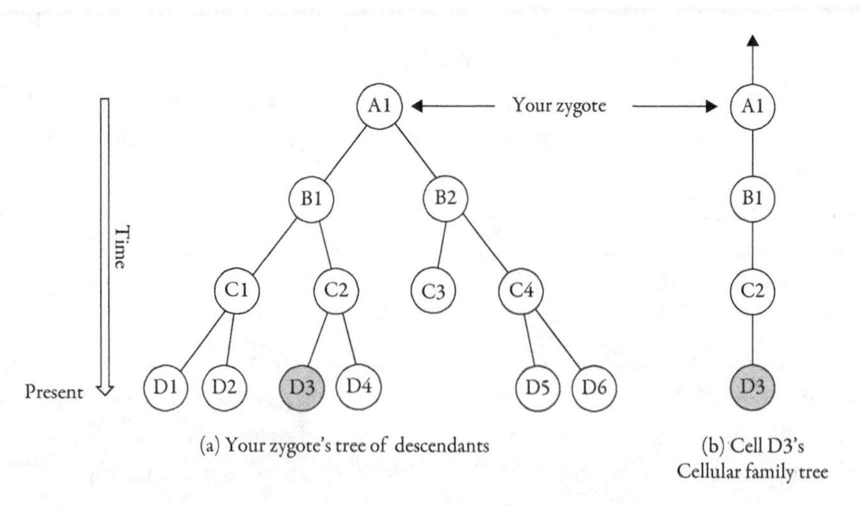

FIGURE 9.1. (a) On the left is your zygote's much-simplified tree of descendants, showing the daughter cells to which cell fission gives rise. Notice that cell C3 died, meaning that its cell line died. (b) On the right is randomly chosen cell D3's cellular family tree, showing its mother cell, its mother's mother cell, and so on. The sister cell of D3 and the sister cells of its ancestors are not shown, since they are not direct ancestors of D3. Since A1 (your zygote) itself had ancestors, D3's cellular family tree would extend beyond what the above diagram shows.

two more branches, if the daughter cells have also fissioned. The exponential nature of cell fission has astonishing consequences. According to one estimate, by the time you became an adult, your zygote had fissioned into 3.7×10^{13}—that's 37 trillion—cells.[7] To get an idea of what a cellular tree of descendants would look like, see Figure 9.1.

The cellular family tree shown in Figure 9.1 is of course incomplete, inasmuch as your zygote itself had cellular ancestors. Depicting these ancestors will require a change in the structure of the tree. Whereas the cells below your zygote came into existence as the result of cell fission, your zygote itself is the result of a cell merger between your mother's egg cell and your father's sperm cell. We can use *T*s to indicate these mergers (see Figure 9.2). And of course, your parents' sperm and egg cells can themselves be traced back to *their* zygotes, which in turn were the result of a merger of your grandparents' egg and sperm cells. And so on, back through your family tree.

Because your cellular family tree will trace back through the zygotes of your ancestors, the resulting tree will be similar to your personal family tree in its overall structure. That said, there will be one very important difference between them. In chapter 2, we saw that it is possible for one individual to occupy multiple positions on your personal family tree. Cells, by way of contrast, can occupy at most one place on a cellular family tree.

Keep going back on your cellular family tree, and you will ultimately arrive at the time before sexual reproduction, meaning that the *T*s indicating cell mergers will stop appearing. What you will instead have is lines that lead from a cell to its mother cell, then to the mother of its mother cell, and so on, all the way back to the cell known as LUCA, the Last Universal Common Ancestor of all currently living things. You will end up with a roughly diamond-shaped diagram, with LUCA as the top vertex, and the cell whose tree it is as the bottom vertex, as is shown in Figure 9.2.

Before we move on, two observations are in order. First, realize that although the cellular family tree of one of your cells will ultimately reach LUCA, the tree will not end there. After all, LUCA itself has a cellular family tree, and that tree will therefore be part of your own extended cellular family tree. Second, it is possible for a cellular family tree to be much more complicated than the ones shown in Figures 9.1 and 9.2. This will be the case, in particular, if a person is a chimera, as was Lydia Fairchild.

Cellular family trees might seem like curious constructs, but this is because you are used to taking persons as the primary biological entity. If your cells could think and talk, though, they would disabuse you of such notions. They would insist that *they* are the primary biological entity and that people like yourself are merely temporary agglomerations of them. Consequently, rather than thinking of them as *your* cells, it makes more sense to think of your body as *their* construct.

HOW OLD ARE YOUR CELLS? How we answer this question depends on what we think happens to a cell's identity when it undergoes cell fission and fusion. A case can be made that such events do not reset a cell's age clock. A cell, after all, remains alive throughout

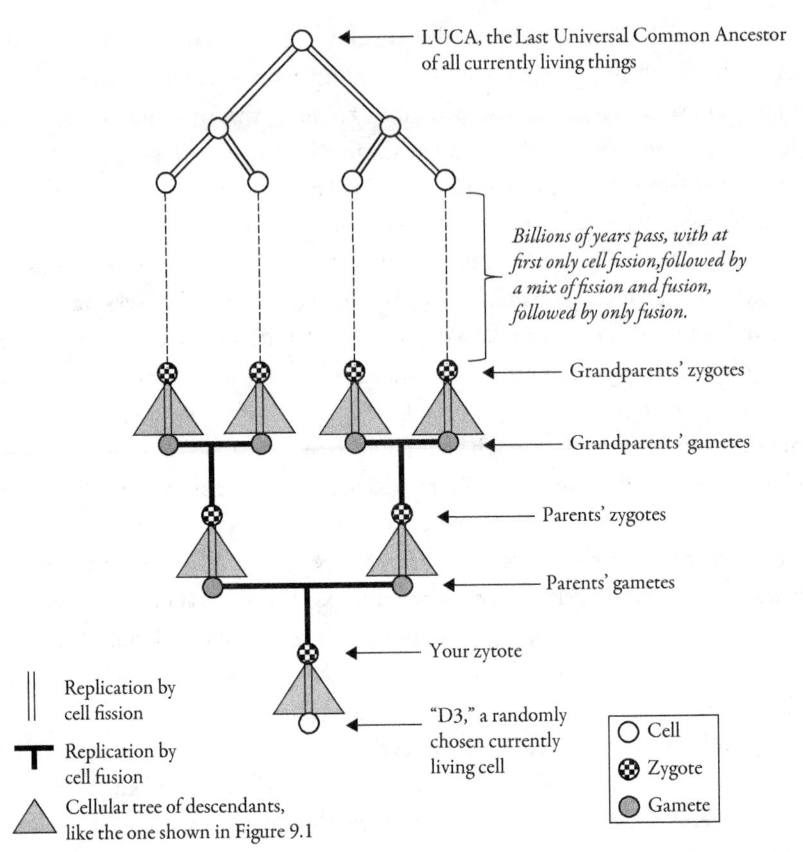

FIGURE 9.2. This is a radically simplified version of the extended cellular family tree of randomly chosen cell D3 of Figure 9.1. Between it and your zygote, there would have been multiple instances of cell fission, but the zygote itself would have been the result of cell fusion. If we go far enough back on this tree, though, we will come to a time before sexual reproduction, when cells divided but didn't fuse. In between those times would have been a transitional period, with some fission and some fusion. This is indicated by the dashed lines in the diagram. As we move up this tree from the current cell, it gets wider, a consequence of cells having come into existence as the result of cell fusion. As we move down from LUCA at the top of this tree, it also gets wider, a consequence of repeated cell fission, first of LUCA and then of LUCA's descendants. The resulting cellular family tree will be shaped like a diamond, wide in the middle and converging onto vertices at the top and bottom. What has taken place to produce this tree is quite remarkable: over billions of years, one cell, LUCA, has become many, which in turn have become one cell, your zygote, which in turn have become the many cells that comprise you.

such events. In fission, one cell simply reorganizes itself into two cells, and in fusion, two cells combine to make one. Measured in this way, a cell's age will be the time that has passed since its oldest cellular ancestor came alive, meaning that the cells that comprise you have existed—and been alive—for billions of years. It is a remarkable achievement. Furthermore, it holds open the possibility of cellular immortality.

Many people, however, will reject the notion of "continuing cell identity." They will instead argue that when a cell undergoes fission, two entirely new cells come into existence, and that when two cells undergo fusion, one entirely new cell comes into existence. They will therefore take a cell's age to be the amount of time that has passed since the fission or fusion event that brought it into existence. In this way of thinking, your blood cells are only a few months old. Your skin cells are also relatively short lived. Your brain cells, though, are decades old; otherwise, you wouldn't be able to remember your childhood. And if you are a woman, your egg cells came into existence before you were born, making them older than you are—if, at any rate, we take your age to be the one shown on your driver's license.[8]

And what, you might wonder, is the average age of your cells, as measured by the amount of time that has passed since the fission events that brought them into existence? One way to find out would be to mark your cells with a radioactive tracer and then measure how the level of radioactivity changes with the passage of time: realize that each time a cell divides, the amount of tracer it carries will drop by half. Scientists have done this sort of experiment on mice but have understandably been reluctant to do it on people. It would be hard to find test subjects willing to become sufficiently radioactive for the experiment to work.

In the 1950s and early 1960s, though, the nuclear arms race created a few billion involuntary human test subjects. Because of above-ground testing of nuclear weapons, the amount of carbon-14 in the environment soared. (Atoms of carbon-14 have two more neutrons than carbon atoms usually do, and the presence of these additional neutrons makes them radioactive.) These carbon-14 atoms were taken in by the people then alive and incorporated into their cells. And because above-ground tests subsequently ended, biologists were able to measure the rate at which the radioactivity of various types of cells declined and thereby infer the life spans of these cells.[9] They concluded that the average cell in an adult's body is only 7 to 10 years old, meaning that even though your driver's license says you are ready for retirement, the cellular you is (on average) old enough to be in grade school.

IN WRITING THIS BOOK, I FIND MYSELF in the position of a philosopher writing about science. As a philosopher, I am attuned to questions about identity. In doing the research for this book, though, I have found that scientists tend to gloss over such questions. Since their primary goal is to formulate and test general theories, it is perfectly understandable that they would do this. My goal in writing this book, however, is to reveal something important about the identity of you, my reader—namely, that you have not one identity but many. For me to accomplish this goal, I must take questions of identity seriously. This is why, back in chapter 4, I wrestled with the concept of the identity of a species. It is a concept that does not have nice, sharp edges, meaning that any attempt to sort living things into species will be rather arbitrary. This is also why, in this chapter, I ask what happens to the identity of a cell when it divides. Does it become two new cells, or is it simply the same cell that has broken into two parts?

To set the stage for some of the questions that will arise as we continue our investigation of who and what you are, let me pause here to introduce what is known as the *ship of Theseus paradox*. It was described by biographer Plutarch in the late first century, and ever since, philosophers have been scratching their heads about the identity question it raises. Here, in its simplest terms, is the paradox. Suppose that to keep a wooden ship seaworthy, you replace its planks when they rot. If you replace only one plank, it clearly remains the same ship, right? But suppose that as a result of a long series of replacements over many years, *all* the planks in the ship have been replaced. Will it be the same ship as it initially was?

If you think it *is* the same ship, consider this twist. Suppose that as you discard the rotten planks, your frugal neighbor takes them at night, as part of a low-budget ship-building project. In the end, he is able to construct an entire ship out of your discarded planks. It is true that his ship, being composed of rotten planks, will not be seaworthy, but it will nevertheless be a ship. The question: if you have the same ship as you started with, then what is the status of the ship your neighbor built? It is composed of exactly the same planks as your ship used to contain, so is it your old ship? And if it is, then what is the identity of your new ship? Or can it be that the "two" ships, although distinct, are one and the same ship?

It is possible to make this same philosophical point with a joke. A man walks into a dealer of historically significant antiques and offers to sell its proprietor the axe that the youthful George Washington used to cut down the cherry tree. The proprietor is astonished: "You mean to tell me that George Washington once held this axe?" The man's reply: "Well, sort of. Because it remained in use after George owned it, the head has been replaced twice and the handle has been replaced three times."

The questions raised by the ship of Theseus paradox arise whenever we take something to be a collection of things. In particular, if you are a collection of cells, and if those cells are constantly being replaced—indeed, if even the zygote that you originally were has been replaced—are you the same person as you used to be? More such questions arise in part III of this book, when I take you to be a collection not of cells but of atoms.

10

Your Ancestors Were Boring

YOUR MOTHER MAY HAVE BEEN a brilliant conversationalist, your father may have been an accomplished poker player, and your grandfather may have been able to pull quarters from behind your ear—how *did* he do that?—but keep going back in your family tree, and you will find ancestors who were profoundly boring.

To see why I say this, suppose you could travel back in time to visit our planet 2.2 billion years ago. The life forms you encountered—some of which would be your direct ancestors[1]—would be single-celled organisms that spent their days floating in water or maybe swimming through it in a random fashion, propelled by flagella. As a result, the world around you would be quite uninteresting—unless you were a geologist who preferred to see landforms unobstructed by vegetation, or a microbiologist who had the foresight to bring along a microscope.

Travel forward in time to 1.8 billion years ago, though, and things would be rather more interesting. There would, in particular, be single-celled eukaryotes. As we have seen, eukaryotes are one of the three domains of living things, the other two being bacteria and archaea. We will have much more to say about eukaryotes and how they came into existence in chapter 11, but for now, suffice it to say that they can do things bacteria and archaea can't, such as crawl by changing their shapes. Given this progress, you might have high hopes for these eukaryotic organisms, but for the next billion years, they remained pretty much at this stage of development. Some biologists therefore refer to this period as the *boring billion*.

One way for cells to increase their impact on the world is to get bigger, but there are physical limits to how big a cell can get. As a cell grows, its volume increases at a faster rate than its surface area, making it increasingly difficult for the cell's metabolic processes, which increase at the same rate as its volume, to get the nutrients they need and expel the waste they produce. Also, getting bigger puts more pressure per unit area on a cell's surface, increasing the chance that it will burst, at which point the cell will lose its ability to do anything.

The solution to this predicament is for cells to remain tiny but do what tiny things everywhere do when they want to have an impact on the world: combine with other tiny things to accomplish jointly what they cannot hope to accomplish individually. Cells can, in other words, band together to become a multicellular organism. This is what our human cells do. It is also what we humans do: by cooperating with other people, we are able to accomplish things that would otherwise have been impossible, such as building Hoover Dam and traveling to the moon.

DURING THE BORING BILLION, most of the organisms that existed were solitary microbes. When they underwent fission, the resulting daughter cells would simply drift away from each other, or if they had flagella, swim away. One exception to this would have been bacteria that, as the result of a mutation, were sticky. They would have stuck to the surfaces they encountered and on reproducing, their daughter cells, which would have inherited their stickiness, would have stuck around as well. A biofilm would result. Such films are still with us; indeed, they are responsible for some of the nastiest infections a person can get.

For further insight into the way multicellular colonies can arise, consider the eukaryotic cells known as *choanoflagellates* (see Figure 10.1). They spend most of their time as solitary cells, floating free, but when certain foods become available, they join together

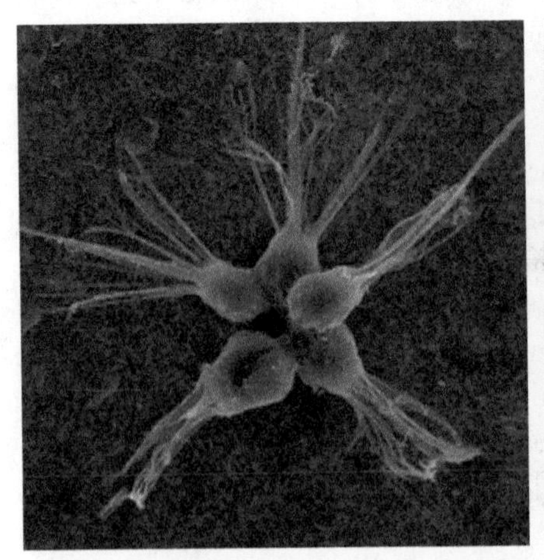

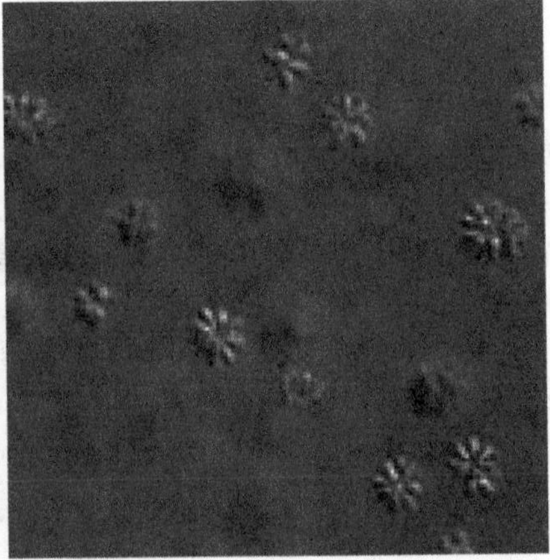

FIGURE 10.1. On the left, choanoflagellates have combined to form a colony. On the right, several such colonies are shown. Creative Commons: http://www.dayel.com/choanoflagellates/.

to make rosette-shaped colonies in order to better consume that food.[2] It is also possible for choanoflagellates to be "born into" colonies. Such colonies are founded when the two daughter cells that result from cell fission are connected by a thread. When those daughter cells subsequently fission, their daughter cells are also connected, and so on. The end result will be a colony of genetically identical cells.

There are, as we have seen, limits to how big a cell can get. There are also limits to how big a choanoflagellate colony can become. When a colony is small, all its cells are "outside" cells, in contact with the environment to which they are accustomed. As it grows, though, there will eventually be "inside" cells that, instead of being exposed to nutrient-bearing water, will be surrounded by other choanoflagellates. At some point, the colony's interior environment will become unlivable. Fortunately, this "inside cell" problem has a solution; otherwise, macroscopic multicellular creatures like ourselves wouldn't exist.

For insight into the nature of this solution, consider sponges. They look like plants, rooted to rocks on the bottom of the ocean, but they are in fact colonies of tiny animals. The walls of a sponge's vase-shaped body have to be thick to support its three-dimensional structure, meaning that the cells inside the sponge's "flesh" will be surrounded by other cells. To take care of the needs of these inside cells, a sponge is riddled with tubes that transport water through its flesh, into its central cavity, and then out the opening at the top of this cavity, as shown in Figure 10.2. The flow of water through these tubes can be revealed by putting dye into the water outside the sponge. In a few seconds, the dye will appear in the central cavity and then will drift up through the top opening of the sponge, making it look like a little chimney giving off iridescent smoke.

Water flows through these tubes because of the incessant beating of the flagella of the cells that line them. These "pump cells" are called *choanocytes* because of their resemblance

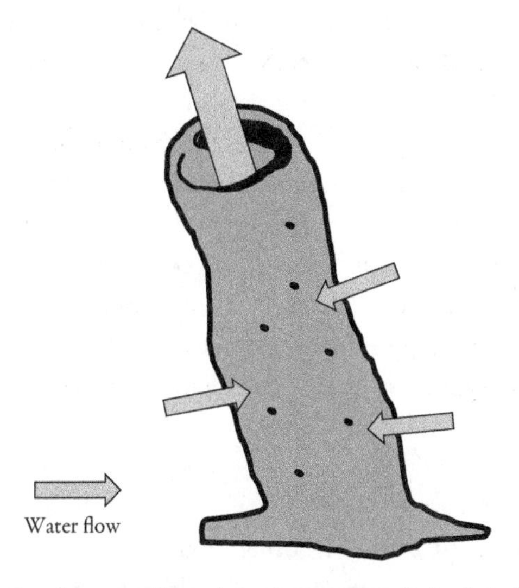

FIGURE 10.2. Nutrient-bearing water flows through the porous flesh of a hollow, cylindrical sponge to meet the needs of the cells that live within that flesh.

to choanoflagellates, a resemblance that apparently is not a coincidence. To the contrary, choanocytes and modern choanoflagellates appear to be close relatives on the tree of life.

If you force a sponge through a very fine sieve and thereby disaggregate its constituent cells, those cells will not only remain alive but will reaggregate to make a new sponge. And if you mix together the disaggregated cells of two different sponge species, they will independently reaggregate into two different sponges. It is as if they *want* to be multicellular, as if they regard multicellularity as their natural state.

Whereas sponges filter-feed through their flesh, sea anemones take in food through their mouth, process it, and then expel waste and undigested food through that same mouth—or should we call it an anus? It is a digestive process that is simultaneously inefficient and unsanitary. As a result, it is quite rare in the animal world.

Instead, the bodies of most animals are built around a tube known as a gut. At one end of this tube is the mouth, where food and water enter, and at the other end is the anus, where waste is excreted. It is a brilliant design. For that gut to work, though, the animal needs a way to move water and food through it. Whereas a sponge moves water and food with flagella, a worm does it with muscles. Muscles also allow a worm to move to where food is, something a sponge is unable to do. Using these muscles consumes energy, of course, but the throughput strategy of a gut is so effective that worms have energy to spare.

Although a tube with a mouth and anus is a very clever design, it is capable of improvement. The earliest worms would have swum or tunneled in a random manner. They relied on luck to get their lunch. Their descendants, though, gained the ability to sense their environment and in particular to sense the presence of certain chemicals in the water. With this information, their motion could be directed toward where food was likely to be. Since the goal of this directed motion was to bring food to the worm's mouth, the logical location for its sensory organs was near that mouth. Later, when worms developed primitive brains, the logical place for them was near the source of the sensory input that they would analyze and therefore was near their mouth as well. This is the evolutionary reason why what we call our head is home not just to our mouth, but to our taste buds, nose, eyes, ears, and brain. It is the design favored by most animals.[3]

In evolutionary terms, we humans represent an advanced stage of the tubular body design: we are, at our core, a tube with a mouth, a digestive tract, and an anus. We inherited this design from a distant worm ancestor. Over millions of generations, we have evolved in a manner that allowed this "inner worm" to better conduct its business. This included the addition of arms, legs, and a head with a brain. It is therefore not too much of a stretch to describe us as radically accessorized worms.

ALLOW ME TO RECAPITULATE. The boring single-celled organisms that roamed the earth 800 million years ago have evolved into multitalented macroscopic multicellular organisms like yourself. For this transformation to have taken place, the "inside-cell problem" had to be solved. Sponges, sea anemones, and worms found different ways to solve it. For any of these solutions to work, though, cells had to specialize.

The cells in a colony of choanoflagellates are minimally specialized, with a given cell performing one of five different functions.[4] Sponge cells differentiate to a greater degree. We have already encountered the choanocytes that specialize in pumping water through the flesh of a sponge. Other cells will form the wall of the sponge, while yet others will control the flow of water through that wall. Worm cells, of course, are even more differentiated than this. As a result, worms can do many things that sponges cannot.

Consider, for example, *Caenorhabditis elegans*, a 1-millimeter-long worm—more precisely, a nematode. Its 1,000 cells are much more differentiated than those of a sponge. It has nerve, muscle, skin, intestine, and reproductive cells,[5] meaning that unlike a sponge, it can sense and move around in its environment. It might even possess a sex drive.[6]

You, of course, are vastly more complex and talented than a worm. Your cells are also more specialized. We humans are comprised of hundreds of different kinds of cells,[7] among which are liver cells, skin cells, and neurons. Furthermore, within the areas of specialization there will be sub-specialties. Photoreceptor cells, for example, are neurons that are capable of detecting light.

As we saw in chapter 9, your cells are all descendants of a single-celled organism. To become part of the macroscopic, multicellular organism that is you, they had to change their ways. To begin with, they had to give up their ability to roam. And having taken up permanent residence within a multicellular organism, they had to do the same job, all day long, every day of the week—cells do not get vacations. In return, they get what is for them a very nice place to live: unless they are skin cells, they live in a climate-controlled environment. Under normal circumstances, they get all the food they can eat, along with all the water and oxygen they need, and their waste products are efficiently hauled away. I should add that the job they are assigned is one that they will be wonderfully suited to do.

Cells, to be sure, don't think about the advantages of combining with other cells to get things done or the sacrifices they have to make to combine in this manner. This is because cells don't think, period. (Although neurons make human thought possible, they themselves are incapable of thought.) What happens is that cells that cooperate are more likely to survive and reproduce than cells that don't, and the cells that reproduce transmit their cooperative nature to their cellular descendants. Because of their "willingness" to cooperate, your cells can live vastly longer than they would "in the wild."

Although most of your cells stay in one place relative to your other cells throughout their lives in order to do their assigned job, there are exceptions. Your skin cells, for example, are normally stationary, but in response to a nearby injury, they will move toward it, in what is literally a gap-filling measure. Your blood cells spend their weeks of life circulating through your blood vessels. If you are a male, your sperm cells, besides being capable of traveling through your reproductive system, can leave your body to travel through the reproductive system of a female, in search of an egg. And your macrophages, which are a type of white blood cell, crawl around inside you looking for biological trash to clean up. If they come across a dead or dying cell, or cellular debris, they ingest it and recycle its chemical compounds.

In their behavior, these macrophages resemble the single-celled eukaryotes from which they and all our other cells evolved. The free-roaming existence that served them then serves us now. And curiously, even our more sedate cells will reveal their ancestral roots if we separate them from their neighbors. Put a prostate cell in a culture dish, and it will start crawling around.[8] It is as if being removed from the group unleashes its primal instincts.

Freedom to move isn't the only thing your cells—most of them—give up. They are also routinely called on to stop dividing. You want, for example, your liver to reach a certain size and then stop growing. Realize, though, that in asking your cells to stop dividing, you are asking them to stop doing the thing that each of their ancestral cells—every single one of them!—has done since the very first ancestral cell came into existence. It is a lot to ask. And yet it is a price your cells are willing to pay to be part of the collection of cells that is you.

There is, however, an even bigger sacrifice that might be expected of your cells. They can be called on to give up their lives for the greater good, in the process known as *apoptosis*. Consider, by way of illustration, the construction of your hands while you were in the womb. At the early stage of their development, your hands resembled not gloves with independent fingers, but mittens with all their fingers joined. Then, five weeks into your fetal development, the signal went out for selected cells—those in what is known as the *digital interspace*—to die, and as a result, independent fingers appeared. If these cells had refused to die, you would have a birth defect.

Occasionally, cells rebel against collective cellular authority. This is what, unfortunately, cancer cells do. At first, they ignore the injunction to stop dividing or to die. Later in their development, they ignore the injunction to stay in one place. They instead revert to their ancestral state: they break loose from their assigned address in the multicellular city and head for distant districts. When this happens, doctors say that the cancer has *metastasized*.

CELLS, AS WE HAVE SEEN, benefit from joining together to make a multicellular organism. The resulting organisms, however, can also join together to make what biologists refer to as a *superorganism*. This is what bees, termites, and corals do.[9] The bees in a hive are not physically connected, the way their cells are, but they are genetically connected in the sense that almost all of them are sisters. This means that a beehive might be thought of as a *multi*-multicellular organism

The bees in a hive have specialized jobs that change as they age. At one point in their lives, worker bees might clean the hive and at another they might construct brood cells. Later, they might spend their days bringing pollen, nectar, or even water back to the hive.[10]

Later still, they might guard the entrance to the hive. And when its working days are over, a bee's body is simply thrown from the hive by hive-cleaning bees. Unlike the queen, these worker bees never reproduce. The queen does that job on their behalf. Then again, we might want to say that it isn't the queen that is reproducing but that the superorganism known as the hive is reproducing through the queen, which might be thought of as the hive's reproductive organ.

Something like this happens in your own body. Your cells are divided into somatic cells and germ cells. The former cells, which comprise almost all of you, ultimately perish without leaving any descendants behind. Like worker bees, they spend their lives doing an assigned task so that something else—namely, your germ cells, which are their genetic twins—can reproduce on their behalf.

THE ZYGOTE THAT YOU USED TO BE was a stem cell.[11] As such, it didn't have a job to do, other than to reproduce by means of cellular fission. Its daughter cells were likewise stem cells. But after about a week of dividing, your cells started to specialize. They didn't do so by choice; instead, they were told, by a chemical signal, what job they would do.

A cell specializes by selectively expressing its genes. Activate some genes, and a cell will play the role of a skin cell; activate other genes, and it will instead become a liver cell. We humans have about 20,000 different protein-coding genes. It has been estimated that 8,847 of them are activated by every one of our cells. For a cell to play the role of a neuron, however, requires the activation of 318 additional genes. By way of contrast, the cells in your testicles—assuming that you are a male—require the activation of 999 additional genes, more than are required by any other cell in the human body.[12]

When a specialized cell divides, the daughter cells not only inherit the mother cell's DNA, but they inherit instructions about which genes in that DNA are to be activated. They thereby inherit the specialization of the mother cell. In the normal course of things,[13] cell specialization is a one-way street: once a cell has specialized, it cannot revert to a stem cell, nor can it change specializations.

For your zygote to give rise to hundreds of different specialized cells, it had to possess, in its DNA, the "blueprint" and "operating manual" for each of these specializations. In other words, your zygote had to "know" how to make and operate a liver cell, a skin cell, a neuron, and so on. It is understandable that the zygote would have to do this, but curiously, even after cells specialize, they continue to carry your complete genome, not just the part that is relevant to their specialization. As a result, a skin cell "knows" how to be a neuron, but since it has been instructed to become a skin cell, it will never put this knowledge to work.

We have seen that for single-celled organisms to transform into multicellular macroscopic organisms, they had to surmount the "inside-cell problem," but this wasn't the only problem they encountered. The more complex a multicellular organism is, the more cell specialization it will require. This will in turn require a bigger genome to carry the blueprints and operating manuals for all those specializations. By way of illustration, the genome of *Salpingoeca rosetta*, a choanoflagellate capable of making colonies, is 55 million base pairs long, the genome of the rather more complex *C. elegans* nematode is 100 million base pairs long, and the human genome is 3.2 billion base pairs long.

Genomes, however, are biologically expensive to maintain and copy, meaning that an increase in the size of an organism's genome would have a significant impact on its energy needs. But where would the additional energy come from? Luckily for our ancestors—and therefore luckily for ourselves as well—a new energy source fell into their lap, as it were. It is to this remarkable event that we turn our attention in the next chapter.

11

Your "Cellmates"

IN THE PREVIOUS CHAPTER, WE EXPLORED the consequences of becoming a multicellular, macroscopic organism. For this to happen, your cells had to specialize, which in turn meant that they needed much bigger genomes. A microbe like *E. coli* has to contain DNA instructions for building and operating one kind of cell. This can be accomplished with a genome only 5 million base pairs long. Your cells, though, have to contain instructions for building and operating hundreds of different types of cells. As a result, your genome has 3.2 billion base pairs, more than 600 times as many as that of an *E. coli*.

Maintaining and copying genomes, however, takes a lot of energy. Until about 2 billion years ago, this energy simply wasn't available to organisms, so they remained small and simple. Then something happened. A new source of energy became available, and as a result, microorganisms were able to evolve into complex, macroscopic organisms capable of engaging in activities that consume lots of energy, such as swimming, walking, flying, and even thinking.

I shall refer to the event that made this energy available as the *Big Gulp*—scientists refer to it as the *endosymbiotic merger that gave rise to mitochondria*—and will spend this chapter explaining what gulped what, how the gulping took place, and how life was subsequently affected. Before I do this, though, some background might be helpful.

EARTH'S LIVING ORGANISMS ARE DIVIDED into three domains: bacteria, archaea, and eukaryotes. You are a eukaryote, as are all other animals, as well as plants, fungi, and algae. One thing eukaryotes have in common is that their DNA is contained in a cell nucleus. This is why they are called *eukaryotes*—from *eu*, meaning *well-formed*, and *karyon*, meaning *nut or kernel*, with the nucleus being the kernel in question. In bacteria and archaea, by way of contrast, DNA floats around freely in the cell. They are therefore referred to as *prokaryotes*—from *pro*, meaning *before*, and *karyon*.

Although it is usually easy to distinguish eukaryotes from prokaryotes—no one would confuse a donkey with an *E. coli* bacterium—distinguishing between prokaryotic bacteria and prokaryotic archaea is rather tricky. Through a microscope they look alike, but if you examine their ribosomes, as microbiologist Carl Woese did in the 1970s, you discover important differences. Ribosomes, it will be remembered, play a key role in deciphering DNA "recipes" in order to make proteins. Woese discovered that archaeal ribosomes work differently than those of bacteria. Consequently, the difference between bacteria and archaea can be likened to the difference between PC and Mac computers: they look alike and do similar things but have different operating systems.

Everyone knows about bacteria. They are all around us. Some of them can make us sick, but they also play a key role in making some of the foods we love—cheese and yogurt, for instance. Archaea, on the other hand, are obscure organisms. This obscurity presumably is a consequence of how little their existence affects our own. Archaea apparently don't make us sick, which is a good thing, since they are resistant to antibiotics. Another reason archaea have such a low profile is that they like to live in places we humans like to avoid—places, for example, that are very salty, hot, acidic, or radioactive.

Consider, by way of illustration, the class of archaea known as *methanogens*. They like to live where there is no oxygen, such as in the muck of swamps or in sewage sludge. This is because the presence of oxygen would deprive them of their favorite "food"—namely, hydrogen. Methanogens make their living, metabolically speaking, by combining carbon dioxide and hydrogen to make methane and water. If oxygen were present, it would tend to combine with any free hydrogen to make water, thereby depriving methanogens of their lunch.

Biologists were slow to accept Woese's discovery,[1] but when they did, they had to restructure the tree of life. Formerly, the tree would have started with a single trunk that divided into two branches, one for prokaryotes and the other for eukaryotes, but since there were three domains of living things, the tree had to have a third major branch. Where should it be placed, though? What, in other words, was the order in which the three domains arose, and how were they related to each other? And most significantly, how did they arise?

WITH THESE QUESTIONS IN MIND, LET US TURN our attention to the story of the Big Gulp. One fine day about 2 billion years ago,[2] two microbes happened to be in the same place at the same time. One was a bacterium and the other was an archaeon. What happened next is the subject of biological debate—indeed, it has been described as "one of the most enigmatic events in the evolution of life on Earth"[3]—but there is general agreement that somehow or other, the bacterium ended up inside the archaeon. This is the "Big Gulp" of which I speak. In the normal course of things, the bacteria either would have been digested by the archaeon or would have killed it, but neither of these things happened. Instead, both the bacterium and the archaeon flourished. Not only that, but their descendants are still around, with the descendants of the archaeon being your cells and the descendants of the bacterium being the mitochondria within those cells. It is these mitochondria that give you the power you need to be the magnificent organism that you are.

To make my story-telling more fluid, I am going to take the liberty of giving proper names to the microbes that are the story's principal characters.[4] I will refer to the post-merger microbe, consisting of a bacterium inside an archaeon, as *Lynn*, in honor of Lynn Margulis, the person most responsible for winning acceptance of symbiogenesis, the theory that puts a scientific foundation under the Big Gulp story. I will refer to the archaeon and bacterium that merged to form Lynn as *Archie* and *Becky*, with Archie being the archeon and Becky being the bacterium (see Figure 11.1). It is thought that Archie was an ancestor of the recently discovered phylum of archaea known as *Lokiarchaeota* and that Becky was an ancestor of modern *Alphaproteobacteria*, a group of microbes that includes parasitic bacteria such as *Wolbachia* and *Rickettsia*.

Archie was arguably a methanogen[5] and as such hated the oxygen that bacteria like Becky typically require. It is therefore strange that these two organisms would be in the same place at the same time, but they were, and Becky somehow ended up inside Archie. It is possible that Archie "swallowed" Becky. Alternatively, Becky might have invaded Archie in an attempt to parasitize her.[6] And finally, Becky could have ended up inside Archie without any "intent" on the part of either microbe: perhaps Archie simply grew around Becky, the way a tree might grow around a nearby fencepost.

Usually when one microbe ends up inside another, their relationship is short lived because one kills the other, but the encounter between Archie and Becky turned out to be the start of a long and beautiful relationship. Becky found the inside of Archie to be a very agreeable environment, so agreeable that she started replicating, making daughter Beckys, who in turn made grand-daughter Beckys. And Archie, rather than being distressed by the presence in her of all these bacteria, profited from it. They provided her with a source of power that enabled her to out-compete rival microbes. Archie would provide Becky's offspring with oxygen and organic compounds, and they in return would engage in highly efficient aerobic respiration that yielded an abundance of energy-bearing adenosine triphosphate (ATP) molecules, some of which they shared with Archie. This arrangement allowed Archie to tap the energy-producing potential of oxygen, as well as extend the range of environments in which an archaeon like her could comfortably live. It was a microbial win-win situation.

Consequently, the merged cell we are calling *Lynn* grew fat and divided. Lynn's daughter cells were "infected" by Becky's offspring: some would have ended up in one daughter cell and some in the other. And when Lynn's daughter cells themselves grew fat

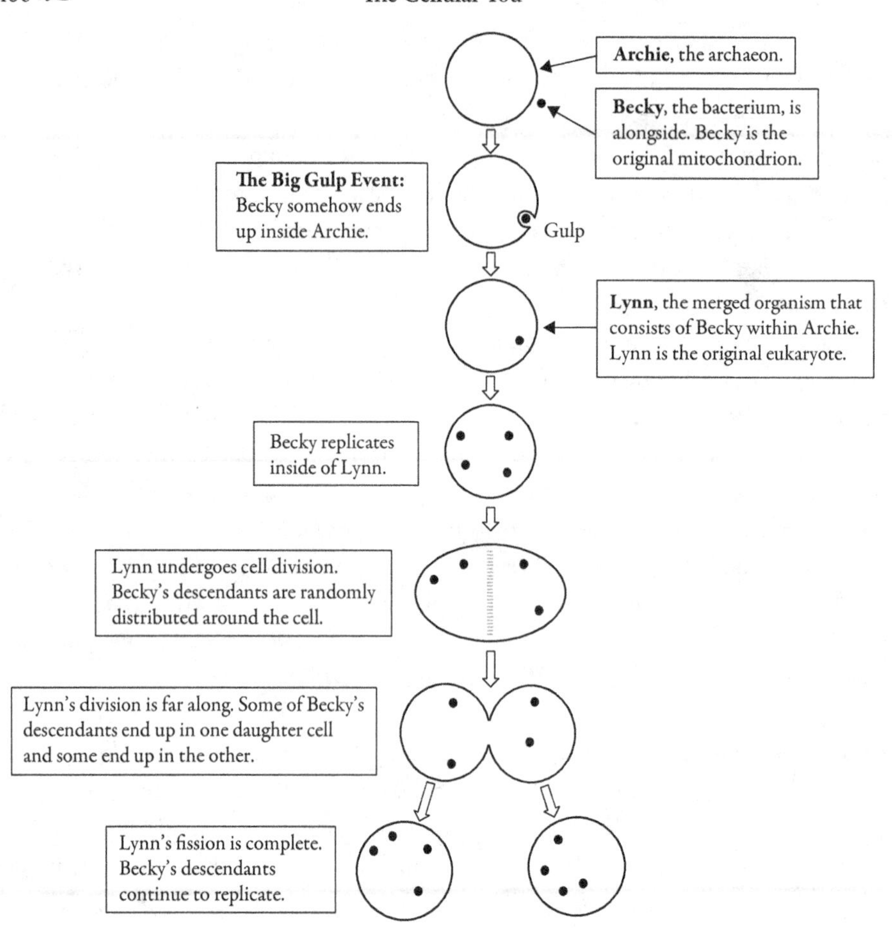

FIGURE 11.1. The (First) Big Gulp event took place about 2 billion years ago. Becky the bacterium is somehow engulfed by Archie the archaeon—biologists disagree on what, exactly, happened—to form the composite organism I am calling Lynn. Becky is so well suited to her new environment that she keeps growing and dividing. And because Archie doesn't mind Becky's presence inside her—or even benefits from it—she also keeps growing and dividing. When Lynn divides, Becky's descendants end up in each of the daughter cells. You have Becky's descendants within your cells. They are your mitochondria.

and divided, *their* daughter cells also carried the descendants of Becky. Your 37 trillion cells are all descendants of Lynn, and therefore of Archie as well. Within almost all of your cells reside the descendants of Becky.[7] As a result, you have cells within your cells—they are your cellmates, as it were—but you probably don't think of them that way. You instead call them your *mitochondria*.[8]

I mentioned earlier that unlike bacteria, archaea are unaffected by antibiotics. This is presumably why your cells, which are the descendants of archaea, are unaffected by the antibiotics you take to deal with bacterial infections. Significantly, though, the mitochondria within your cells can be negatively impacted by antibiotics, a susceptibility that is presumably a consequence of the evolutionary past of these organelles.[9]

In chapter 9, we explored your cellular family tree. We saw that all your current cells can trace their ancestry back to the one cell that was your zygote, and that all the cells that currently live on Earth can trace their ancestry back to the cellular organism known as *LUCA*, the Last Universal Common Ancestor.[10] According to the Big Gulp theory, between these two "choke points" in your cellular ancestry, there was a third. It was occupied by the organism I am calling Lynn (see Figure 11.3). And not only can each of your cells trace its ancestry back to her; every cell of every eukaryote can also trace its ancestry back to Lynn. This includes the cells of the spider in the corner of your bedroom closet as well as the cells of the maple tree in your back yard. It includes the yeast cells in the bread dough rising in your kitchen and the fungal cells that have invaded your big toe. This is quite a posterity for one cell to have.

YOU INHERITED YOUR MITOCHONDRIA exclusively from your mother.[11] They are all descendants of the mitochondria that were in the egg that, on being fertilized, became your zygote. Your father's sperm also carried mitochondria, but the first thing that happened to it on entering the egg was that it lost its head. More precisely, its head entered the egg, while its tail and the midpiece that propelled that tail remained outside. Since a sperm's mitochondria are in the midpiece, your father's mitochondria normally wouldn't have gained access to the egg.

How many mitochondria a cell has depends on its energy needs. An energy-hungry heart muscle cell might have 6,000 of them, enough to constitute 40 percent of its cytoplasmic volume.[12] A liver cell might have 1,000–2,000 mitochondria, enough to constitute 20 percent of its cytoplasm.[13] Sperm cells might have only 100 mitochondria, and red blood cells have none. Furthermore, how many mitochondria a cell has will vary in accordance with the recent energy needs of that cell. An athlete in training will have far more mitochondria in her muscle cells than will a sedentary person.

According to one estimate, the cells of adult humans will have, on average, 300–400 mitochondria, enough to constitute 10 percent of our body weight.[14] And of course, for that many mitochondria to fit into a cell, they must be very small, and indeed they are, as their bacterial ancestor Becky doubtless was.

When they first met, Archie and Becky had their own DNA, and on merging, this DNA did not "combine," the way the DNA of an egg and sperm do. Instead, Becky's DNA was separated from Archie's by Becky's cell walls. Consequently, Lynn carried two different sets of DNA, and because you are a descendant of Lynn, you do as well. Within the nuclei of your cells, you have what biologists refer to as *nuclear DNA*. It is derived from the DNA of Archie, and you likely refer to it as *your* DNA, inasmuch as it is what gives you your five fingers, your blood type, and the color of your eyes. Your mitochondria, though, also contain DNA. This *mitochondrial DNA* (or *mtDNA*) is derived from that of Becky and is strikingly different from your nuclear DNA. Whereas the chromosomes in your nuclei are linear, those in your mitochondria are circular, the way bacterial chromosomes are. Not only that, but as we saw in chapter 7, the DNA of your mitochondria is deciphered using a slightly different genetic code than your nuclear DNA. People talk about "their" genome, but this is misleading, since they in fact have two different genomes, one for their cells and another for their mitochondria.

In chapter 9, we encountered human chimeras that contain within them the body parts of former womb-mates. We also saw that most people are microchimeras: among their cells lurk cells of their mother or, if they are mothers, cells of their offspring. But it should now be clear that we are chimeric in a more radical sense of the word: our cells carry the DNA of two different ancient organisms. The organisms in question were not different members of the same species, as is the case with human chimeras and microchimeras. They instead belonged to two different species, which in turn belonged to two different domains: one was an archaeon and the other was a bacterium. This makes you a cross-domain, intracellular nanochimera. Congratulations are clearly in order.

Although your mtDNA is derived from that of Becky, it has been transformed by the passage of time. Because Archie's descendants did such a good job of providing for their needs, Becky's descendants started losing genes: if Archie's descendants were going to provide them with a certain protein, why make it themselves? And if you aren't going to make a protein anymore, why keep the gene recipe for making it? As the result of 2 billion years of this gene-paring process, Becky's modern descendants—your mitochondria—are down to only 37 genes.

At the same time as the mitochondrial DNA of Lynn's descendants was shrinking, their nuclear DNA was getting bigger. Whereas Archie's nuclear DNA might have had a few million base pairs, the way modern archaea do, her eukaryotic descendants—including your own cells—can have billions of base pairs.[15] This DNA growth was made possible by the power provided by Becky's descendants. Thanks to a bigger genome, Lynn's descendants could engage in the cell specialization that, as we have seen, is necessary to become a multicellular macroscopic organism. You should therefore embrace your nanochimerical nature. Without it, you would likely be as boring a single-celled organism as Archie must have been.

It is somewhat surprising that our "human" cells would be descendants of an archaeon. Archaea, after all, tend to live in extreme environments, in places we humans would find too salty, hot, acidic, radioactive—or all of the above.[16] Nevertheless, our cells are more closely related to the archaea happily ensconced in a scalding hot geyser than they are to the bacteria in the yogurt that we call lunch. How very strange!

THE BIG GULP, IT TURNS OUT, wasn't the only instance of an endosymbiotic merger in which one organism engulfed another and the two organisms lived happily ever after. Indeed, there is evidence that Archie herself was a chimera, the result of past encounters between her ancestors and bacteria.[17] Furthermore, one of Lynn's descendants—I shall refer to her as *Lynette*—was subsequently involved in another very significant interspecies merger. I shall refer to it as the *Second* Big Gulp and let us, to avoid confusion, start referring to the merger that produced Lynn as the *First* Big Gulp.

As a descendant of Lynn, Lynette would have carried mitochondria. On one fine day, a few hundred million years after the First Big Gulp,[18] Lynette found herself alongside the cyanobacterium that I shall refer to as *Cynthia*, and the old story repeated itself: Cynthia was somehow engulfed by Lynette. Cynthia liked her new home, and as a result started growing and dividing, as did the resulting daughter cells (see Figure 11.2). And because cyanobacteria are capable of photosynthesis, Cynthia provided Lynette with a new source of power, derived from sunlight. It was another win-win situation, and as a result, the descendants of Lynette are still around: we call them *plants* and *algae*.[19] The descendants of Cynthia are also still around: they are the chloroplasts in those plants and algae.

We animals, as I have said, are blessed to have mitochondria. Plants and algae, though, are doubly blessed: they have both mitochondria and chloroplasts. What lucky organisms! Furthermore, plants and algae, besides having the nuclear and mitochondrial DNA that animals do, have a third set of DNA as well. It is carried in their chloroplasts. And one more comment is in order: although we humans, as we have seen, use slightly different genetic codes to decipher our nuclear and mitochondrial DNA, plants and algae don't. They instead use the "universal" genetic code to decipher all three of their DNA sets.[20]

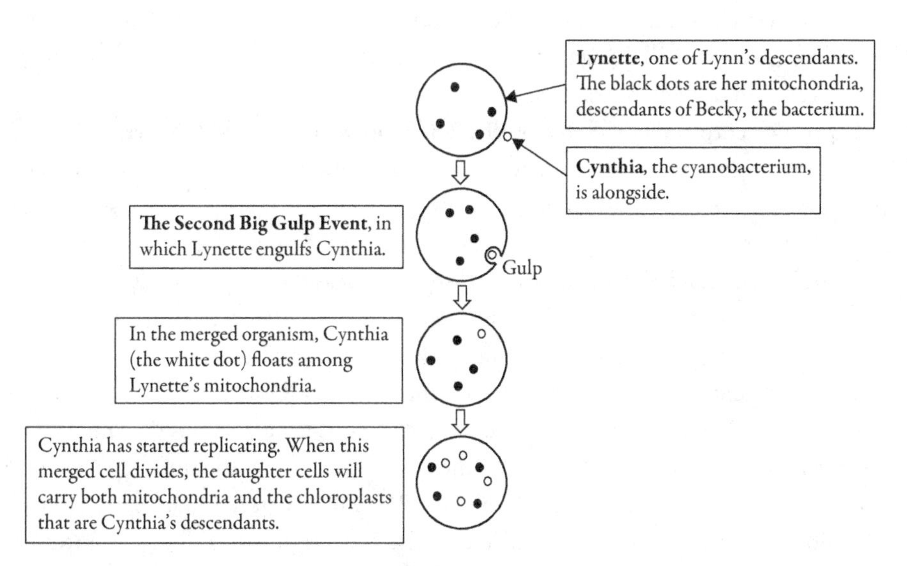

FIGURE 11.2. The Second Big Gulp event took place a few hundred million years after the first. In this event, Lynette—a mitochondria-bearing descendant of Lynn—somehow engulfed Cynthia, the cyanobacterium. The cells of all currently existing plants and algae can trace their ancestry back to the resulting merged cell. It is because plants and algae contain Cynthia's descendants—now referred to as chloroplasts—that they are able to transform light into chemical energy.

We animals are direct beneficiaries of the First Big Gulp. We also benefit from the Second Big Gulp, but indirectly. If it hadn't taken place, plants and algae wouldn't exist, and this would have had a profound impact on animals' subsequent development. For one thing, we would not have had plants to eat—and therefore would not have had plant-eating animals to eat, either. Furthermore, without plants and algae, there would be significantly less oxygen for animals to breathe.[21] Without abundant food and oxygen, though, early animals would have lacked the power necessary to evolve into the complex organisms that they subsequently became. Consequently, we humans owe our existence to the biological equivalent of lightning striking our single-celled ancestors and their cousins not once but twice, in the two Big Gulp events described earlier.

That Archie and her offspring were able to participate in multiple endosymbiotic mergers suggests that she was a talented hostess, so to speak. To fully appreciate this talent, though, there is one other thing we should consider. A few hundred million years after the First Big Gulp,[22] Archie's eukaryotic descendants figured out how to engage in sexual reproduction, an event in which one cell engulfs another. More precisely, an egg cell engulfs a sperm cell. Of course, in these mergers, the engulfed sperm does not carry on a new life of growing and dividing within the egg. Instead, its DNA merges with that of the egg that engulfed it and a new genetic identity thereby comes into existence.

IT IS TIME TO REVISIT YOUR CELLULAR FAMILY TREE. All of your cells are descendants of your zygote. That cell is in turn a descendant of Lynn, which is comprised of Archie and Becky, which in turn can trace their ancestry back to LUCA, the Last Universal Common Ancestor of all living things. LUCA would have been a single-celled organism that reproduced through fission (see Figure 11.3).

It is also time to revisit the tree of life. The trees we constructed in chapter 4 were based on the assumption that although one species can branch into two, two species can't merge into one. Our discussion of Big Gulp events, though, makes it clear that species can in some sense merge; indeed, all eukaryotic species are the result of a "merger" between an archaeal species and a bacterial species. (More precisely, they are the result of a *physical* merger between *members* of two species.) A completely accurate tree of life would indicate this sort of merger. It might, for example, show a twig leaving one species branch and colliding with another, thereby giving rise to a new twig on the tree.

In chapter 6, I argued that trees of life can be unrealistic inasmuch as they imply that one species instantly transforms into another, when in fact the evolutionary processes that give rise to species take thousands of years. Big Gulp events are an exception to this rule. More precisely, a case can be made that the First Big Gulp, although it involved only two individual cells and might have transpired in under an hour, gave rise to a new species. The same can be said of the Second Big Gulp.

Had biologists been present at the First Big Gulp, though, it probably wouldn't occur to them that Lynn was the first member of a new species. Indeed, they would not even

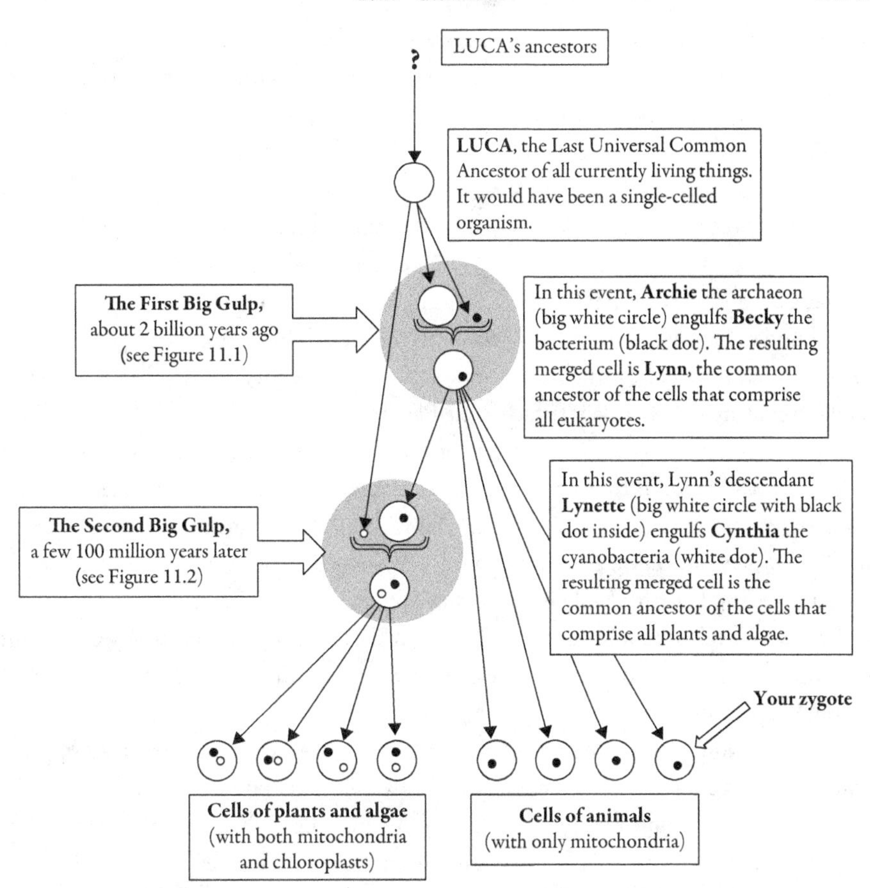

FIGURE 11.3. This is a much-simplified version of LUCA's cellular tree of descendants, over billions of years, down to the present day. Your cells' ancestry can be traced back to your zygote, then back to Lynn, and then (through Archie and Becky) back to LUCA, which itself would have had cellular ancestors. A plant cell's ancestry, however, can first be traced back to the cell resulting from the merger of Lynette and Cynthia. Lynette's cellular family tree can then be traced back through Lynn to LUCA, whereas Cynthia's can be traced directly to LUCA.

count her as a single organism; she would instead be an organism that had another organism inside her. But if these same biologists diligently traced back the ancestry of any modern eukaryote, they might, on arriving at the First Big Gulp, conclude that it represented an obvious place to draw a line of species demarcation—to declare that Lynn was not only a single organism but was of a different species than either Archie or Becky.

In chapter 4, I said that if you went back in time, there would, at any given point, be exactly one species that was your direct ancestral species, and that some of the members of that species would be your direct ancestors. In light of the earlier discussions, some qualifications are in order. You can, as I've said, trace your ancestry back to Lynn. At that point in time, her species would be your only ancestral species, and she would be your one and only direct ancestor. But if you go back to just before the First Big Gulp, you will have not one but two ancestral species, those of Archie and Becky. You will also have those two organisms as your only direct ancestors. As a result, your extended family tree will be interestingly complex. If you were a plant, though, your tree would be more complicated still, since you would also have to take account of the Second Big Gulp.

IT SHOULD BE CLEAR THAT WE OWE A LOT to the humble microbes I have christened Archie, Becky, and Cynthia. Because of Archie's ability to carry other microbes within her, she could host the bacterium that became the mitochondria that, by providing us with abundant power, allowed us to become the macroscopic, complex organisms that we are. She could also host the bacterium that became the chloroplasts that, by allowing plants to capture the energy of sunlight, enabled them to provide us with food. And yet, most of us are oblivious to the existence of these microbes and the role they played in making it possible for us to exist. Some consciousness raising is clearly in order.

I therefore propose that we have an annual Big Gulp Day, during which we celebrate the Big Gulp events. In defense of this proposal, consider the events we do celebrate—for example, the Fourth of July in the United States, Bastille Day in France, and Jesus's birthday in Christian countries around the world. These are all significant events and therefore worthy of celebration, but realize that none of them would have happened if the Big Gulps hadn't taken place.

Ideally, we would celebrate Big Gulp Day on the anniversary of the First Big Gulp, but although this event likely took place in a single day, we don't know what day it was. (The irony is delicious: no one recorded the date of the First Big Gulp because until that event took place, there couldn't have been anyone around to record it.) Let us therefore observe it on March 5, Lynn Margulis's birthday. It was she, after all, who worked so hard to make us aware of this pivotal ancestral event.

12

Your "Boarders"

IN THE PREVIOUS CHAPTER, WE TOOK our exploration of the cellular you a level deeper by considering your "cellmates." These are the "non-human" cells that live within your cells—namely, the bacterial descendants that we call *mitochondria*. Let us now turn our attention to what might be thought of, in cellular terms, as your "boarders." These are the non-human cells that live on and in you, but outside your human cells. The cells in question comprise what biologists refer to as your *microbiome*. It consists mostly of bacteria and archaea, but it can include unicellular eukaryotic microorganisms such as yeast, fungi, and protozoa.

No matter how often you shower your body and wash your hands, your skin is covered with microbes. They tend to be "regional." Some love your forehead because of the tasty oils that are plentiful there. Others reside in the crook of your arm, in your belly button, or on your feet. But we should not make the mistake of thinking that only one kind of microbe lives in one particular region. Rather, microbes of different kinds live in mixed communities, with one kind being more common in one region than in another.

Your pubic hair is also alive with bacteria. They are likely different from the bacteria in your head hair, for the simple reason that these two regions will have different "climates." Furthermore, different people will typically have different pubic microbiomes, a fact that can potentially play a role in courts of law. Suppose, for example, that a rapist, because he used a condom, left no sperm with which he could be identified but left one or more pubic hairs. It is conceivable that the microbiome present on those hairs could play a role in identifying him.[1]

The pores of your skin have microbiomes, which can give rise to acne. Even if you are perfectly healthy, your lungs have a microbiome.[2] The surface of your eyeball also has one, as do your various orifices, including your mouth, ears, nose, urethra, and anus. If you are a female, your vagina has a microbiome not only when you have a yeast infection but when you are "healthy."

Prevent people from washing for a few days and deprive them of deodorants and perfumes, and they will start to smell. How they smell depends to a considerable extent on what microbes inhabit them. Furthermore, because different microbes favor different parts of the body, each part will have a distinctive smell. Your armpits will smell different from your pubic region, which in turn will smell different from your belly button or feet. It is, by the way, no coincidence that some cheeses smell like feet: similar bacteria are responsible for both smells.

WE MIGHT NOT BE SURPRISED by the presence of microbes on our skin—it is, after all, constantly exposed to them—but what about our intestines? They aren't directly exposed to the outside world, and the foods we put in them are either washed, cooked, or both. Furthermore, any microbes that survive this washing and cooking will have to survive the highly acidic environment of our stomach before they arrive at our intestines. We might therefore expect our guts to be microbe-free, or nearly so. And on top of these theoretical arguments, we have the evidence of our own health: doesn't the fact that we aren't vomiting and experiencing diarrhea prove that our guts are microbe-free?

It turns out that our intestines are chock-full of bacteria and archaea.[3] Washing and cooking our food eliminates many but not all of the microbes they carry. And yes, stomach acid kills many of the remaining microbes but not all of them. In particular, *Helicobacter pylori* bacteria are sufficiently tolerant to acid that they regard your stomach not as a gauntlet to be run but as a splendid place to live. And even acid-intolerant microbes can run the gauntlet of your stomach if conditions are right. Suppose, in particular, that when you are dehydrated and your stomach is empty, you drink a large quantity of microbe-laden water. It will pass into your intestines without being acidified. And one other thing to remember: if even one microbe can beat the odds and make it into your intestines alive, it can, in a very short time, give rise to millions of descendants.

Some of the microbes that make it to your small intestine will find it to be an inhospitable environment, but many more will find it to be a microbial paradise. It is a warm, wet place, with plenty of food. You like to eat beans? So do some of the microbes in your gut. As a result of their meal, they will produce gas as a byproduct, meaning that you will experience flatulence. One way to gain insight into your gut biome is to ignite the resulting gas. If it burns with a blue flame, it is evidence that you have methane-producing archaea in your gut. Needless to say, this is an experiment that can go tragically awry, so don't try it at home!

Many of your intestinal microbes will subsequently end up in your feces. Indeed, dehydrate your fecal matter, and between one-quarter and one-half of what remains will consist of the corpses of former microbial residents.[4] It is therefore an understatement to say that you have *some* microbes in your intestines; it is in fact where most of your microbiome resides.

And just how big is your microbiome? Do a census of the cells in or on your body—not including those *inside* your cells—and you will find that for each of your "human" cells—for each cell, that is, that has your nuclear DNA—you have ten cells that are part of your microbiome.[5] Even this remarkable claim doesn't get to the bottom of things, though. It turns out that the bacteria that comprise most of your gut microbiome have resident viruses. Research indicates that these viruses play an important role in keeping their host bacteria healthy,[6] thereby keeping you healthy. It would be misleading, though, to refer to these viruses as part of your microbiome. Your microbiome includes the living organisms that inhabit you—hence the root word *bio*. The viruses that inhabit your resident bacteria are not alive, so biologists instead refer to the collection of them as your *phageome*.

And to be complete, we should keep in mind that the single-celled organisms mentioned earlier are not the only non-human organisms that can live in and on your body. There are also multicellular organisms, including parasitic worms that can inhabit your intestines, lice that can inhabit your hair, and mites that can inhabit your eyelashes.

FOR A LONG TIME, RESEARCHERS ASSUMED that your gut was where the internal portion of your microbiome resided. For this reason, they didn't look elsewhere inside of you for microbes: they "knew" there would be none to find. All this changed rather dramatically in the early 2010s, when researchers stumbled across microbes in all sorts of unexpected places. Even a perfectly healthy bladder, they discovered, will likely harbor microbes. Uteruses, long thought to be germ-free, also harbor them.[7] Discoveries like these triggered a microbial gold rush, as scientists around the globe tried to beat the competition in discovering microbes in unexpected organs.

But if microbes cause illness and if we are full of microbes, then why aren't we sick? Because contrary to common belief, very few microbes are pathogenic. We have seen that archaea don't make us sick, and it has been estimated that only 0.36 percent of bacterial species do.[8] It would appear that we are biased against microbes: because some of them make us sick, we find it easy to believe that they all do.

Along these lines, consider coliform bacteria, the most famous of which is *E. coli*. Feces is typically full of these bacteria, so we should fear them, right? Not necessarily. To begin with, most strains of *E. coli* are harmless. Indeed, the reason they are commonly found in feces is that they live in the intestines of nearly everyone—without, I should add, making their host sick. Coliform bacteria have gotten a bad name in large part because they are used by health officials as an indicator of whether food and water are contaminated.

They were chosen for this role because they are easy to culture. Thus, from the mere fact that tests reveal the presence of coliform bacteria in drinking water, it does not necessarily follow that drinking that water will make you sick. It means only that bacteria that might have a fecal source are somehow ending up in that water, which is cause for concern, since pathogenic bacteria might be among them.

AS I HAVE SAID, UTERUSES ARE NOT microbe-free. This means that babies will have already acquired part of their microbiome before they enter the birth canal. On entering it, they will encounter the microbes that live in the mother's vagina. And once they pop their head out of that vagina, they will, if they are in the usual birth position, find themselves facing their mother's anus, a source of yet other microbes. Babies delivered via C-section don't acquire these microbes, and some have suggested that they would benefit from being "artificially" exposed to them.[9]

A baby's birth microbiome is enhanced when it drinks from its mother's microbe-laden nipples. The milk it drinks, by the way, has evolved not just to feed the baby but to nourish the microbes in the baby's intestines.[10] The baby will subsequently supplement these with other, more foreign microbes by sucking their toys, their thumbs, and maybe even their toes. Before long, the baby will be as full of microbes as its mother.

Over the course of its life, the baby's diet will affect its gut microbiome. This is because microbes, like people, have food preferences. Eating foods high in fat or drinking lots of alcohol will benefit some species of microbe at the expense of others. Similarly, if we eat mostly meat, we will end up with a different gut microbiome than if we eat mostly vegetables.[11] Not only that, but there is reason to think that a change in your microbiome can affect how much and what you eat.[12] It therefore makes sense, when you are deciding what to eat, to keep in mind that you are eating not just for yourself but for the billions of microbes in your gut.

Doctors and dieticians recommend that we include fiber in our diet. Given that fiber is by definition indigestible, this advice raises an obvious question: why should we eat something we can't digest? Because, they explain, the presence of fiber in our intestines promotes the movement of food through our digestive system. There is, however, a second, less obvious benefit. Even though we cannot digest the complex carbohydrates in fiber, many of the microbes in our gut biome can, and we benefit from their presence within us.[13] Eating a high-fiber diet, then, will result in you having a more diverse and more medically desirable gut biome, and will do so without making you fat.

It is possible to change your gut microbiome by radically changing your diet or by taking oral antibiotics. It probably isn't possible, though, to change it in any significant manner by adding probiotics to your diet—by making a point, for example, of having some yogurt with your lunch. People who do this might imagine that by consuming beneficial bacteria they are inoculating what are essentially sterile intestines with good bacteria, and that by doing so they can prevent a takeover by bad bacteria. If they are healthy, though, their intestines will already be densely populated by a variety of bacteria, meaning that any added bacteria will have a hard time getting established.

Your microbiome, it should be clear, plays a huge role in keeping you healthy. Lose it, and you would be in big trouble. This has led some to suggest that it isn't too much of a stretch to think of your microbiome as a bodily organ, comprised of billions of non-human cells.[14]

Your death will have a profound impact on your microbiome. Once you stop feeding them, many of the microbes that formerly flourished within you will perish. Others, though, will start feeding on your dead human cells and thus nourished, will take over your innards: your *micro*biome will thereby be supplanted by a *necro*biome.[15] By analyzing the necrobiome of a rotting corpse, forensic scientists can determine the time of its death.[16]

THE EARLIER DISCUSSION MIGHT HAVE TRIGGERED the "ick factor" in some readers. I have described your body as being permeated with microbes, and perhaps more disturbing still, I have described your gut as being "chock-full" of them. We experience this ick factor because we tend to think of microbes as enemies.

As a result, we might wash our hands with antiseptic soap. Regular soap would do about as good a job of cleaning them, but antiseptic soap makes us feel safer. The problem is that if enough people follow our example, wastewater ends up so full of antiseptics, such as triclosan, that it destabilizes the microbial communities sewage treatment plants use to turn sewage solids into fertilizer.[17]

Because of the ick factor, we might also be tempted to raise our children in as sterile an environment as we can. According to the "hygiene hypothesis," this has the counterintuitive consequence of making them less healthy later in life. Because their immune systems have been pampered, they cannot deal with the microbes and environmental irritants to which everyday life exposes them.

Our dislike of microbes might also cause us, on the first sign of having something wrong with us, to insist that our doctor give us some antibiotics in case it's something bacterial. As a result of such use—and even worse, the use of antibiotics in animal feed to stimulate weight gain—the antibiotics we have discovered are losing their ability to cure us. Disease-causing bacteria have adapted to the antibiotic-filled environment we have created for them, and as a result, they are unaffected by antibiotics that would have killed their ancestors. There is every reason to think that we are about to return to the pre-antibiotic era, in which even a small cut could, on getting infected, take a person's life.

Repeatedly taking antibiotics can devastate a person's microbiome, allowing it subsequently to be taken over by bad bacteria, one of which is *Clostridium difficile*. Its presence in someone's intestines might so disrupt her digestion that she has constant diarrhea—a case so severe that she must wear diapers and must even use a wheelchair, since to stand up is to defecate. And if the *C. difficile* is antibiotic-resistant, as is often the case, the diarrhea in question cannot be cured with an antibiotic. The phrase "death by diarrhea" sounds funny, but it is a rather dreadful medical possibility.

Doctors have discovered that one way to cure such patients is to alter their gut microbiome. They give the patient a dose of feces, taken from a donor with a healthy gut microbiome. The transplant is generally done through the anus of the patient, by means of a colonoscope. Such fecal transplants have been remarkably successful, and have thereby gone from being experimental to being commonplace.

In the process of giving these transplants, doctors have made some interesting discoveries. In one case, a fecal transplant seems to have reversed a lifelong case of alopecia. In another, a 32-year-old woman with a *C. difficile* infection got a fecal transplant in which her borderline-obese teenage daughter served as donor. The mother, who had been of normal weight before the transplant, subsequently started gaining weight, and no matter how much she dieted and exercised, she could not shed the added pounds.[18] Researchers have since done experiments in which they caused weight loss in mice by giving them fecal transplants in which lean humans acted as donors.[19] Cases like these suggest that our gut microbiome has a much greater impact on our well-being than we might have thought. More generally, the microbes in our gut play a significant role in making us the person we are.

WE AREN'T ALONE IN HAVING A GUT BIOME. Koalas also have one; otherwise, they wouldn't be able to digest the eucalyptus leaves they are so fond of. And to ensure that their offspring will be able to digest these leaves, koala mothers periodically give their offspring some of their feces to eat—an orally administered fecal transplant. Ruminants, such as cows, sheep, and goats, also have gut microbiomes; otherwise, they would be unable to digest grass. Termites have a gut microbiome, populated by the protozoan *Trichonympha* and other microbes, that enables them to digest wood. These microbes in turn have a microbiome: the only reason *Trichonympha* can digest the cellulose in wood is because bacteria that live inside it provide it with the enzyme cellulase. When a termite egg hatches, the new termite is inoculated with the gut microbes it will need to digest wood.

Sponges also have microbiomes, populated by bacteria, archaea, and unicellular eukaryotes. These organisms might make up one-third of a sponge's weight.[20] The fact that sponges have about the same body plan as they did 600 million years ago suggests that animals have had microbiomes for a very long time.

And it isn't just animals that have microbiomes. Trees also have them, with different species having different microbes.[21] Legumes have them, in the form of nitrogen-fixing bacteria in the nodules on their roots. Mung beans not only have them but transmit beneficial microbes to their offspring by including them *inside* the seeds they make.[22]

Microbiomes clearly play a role in evolution. An organism that can find and take advantage of beneficial relationships with microbes will be more likely to survive and reproduce than one that can't. But there appears to be another, deeper role microbiomes play. The wasps *Nasonia giraulti* and *N. longicornis* can mate with their distant relative *N. vitripennis*, but the resulting hybrid offspring will usually die. It was assumed that this was because of genetic differences between these wasps, but when researchers used antibiotics to eliminate the wasps' microbiomes, the barrier to reproduction disappeared, and the wasp offspring flourished. When the microbiome was restored, the species again became reproductively incompatible.[23] This suggests that organisms' microbiomes can affect their reproductive compatibility, but not everyone is willing to draw this conclusion.[24]

WE HAVE SEEN WHAT WOULD HAPPEN if you suddenly lost the "non-human" cells *within* your cells: without your mitochondria you would experience a major power outage and would very soon be dead. What about the non-human cells *outside* your cells, though? What would happen if your microbiome suddenly disappeared? Let us divide our discussion of this question into two scenarios.

In the first, we imagine that the only microbes in the world to disappear are the ones now in and on you; the rest remain in existence. From the microbial point of view, your body would become virgin territory in which microbes could live without having to deal with competitors. You would therefore experience the microbial equivalent of a land rush, and many of the microbes that settled on and in you would be nasty ones. There is a good chance that you would survive, but it is also likely that you would initially be very sick.

In the second scenario we shall consider, *all* microbes disappear, including those in, on, and outside of you. The result would be what biologists refer to as a *gnotobiotic world*.[25] You would not, under these circumstances, have to worry about being taken over by nasty microbes. In particular, you would no longer have to worry about bacterial infections. You would, however, have other things to worry about. Without a gut microbiome, you would notice that you were digesting food differently—but still digesting it. Before long, though, you would have to change your diet. For one thing, without bacteria, it would be impossible to make cheese and yogurt. Once the old stocks of these foods had been consumed, that would be it. And the milk used to make them would also disappear, since the cows, sheep, and goats that produce it are ruminants and are therefore quite dependent on their gut biome to help them digest the foods they eat. And of course, without this milk, they would not be able to raise the next generation of cows, sheep, and goats. Conclusion: vegans would find a gnotobiotic world a much easier place to adjust to than carnivores.

But this is only the beginning of the challenges you would face in a world without microbes, since besides needing food to eat, we humans need oxygen to breathe. While it is true that plants are a source of oxygen, cyanobacteria are perhaps its primary source, and in a gnotobiotic world, they would no longer exist. Consequently, it is likely that in the end, even vegans would find it challenging to live in a gnotobiotic world.

Science fiction writers have tried to imagine what it would be like for humans to populate another planet. It would be a difficult undertaking, but suppose Earth were about to get struck by a giant asteroid, or suppose we managed, as the result of pollution or warfare, to make our planet uninhabitable. Humans might, under these circumstances, bravely set forth in search of a new home—and given the discussion in chapter 8, hopefully to someplace nearby. One question that arises is what these interplanetary pioneers should bring with them.

They would, of course, need sufficient food, water, and air for the trip, but they would also need food to eat after they arrived at their destination, and their food supply would have to grow to keep up with what we hope would be a growing population of humans. They could do this by bringing seeds with them and—unless they were willing to embrace a vegan existence—by bringing animals as well. But one other thing they would want to do is bring along the microbes that their plants and animals, and the travelers themselves, would need in order to thrive. Ignore microbiomes, and their colony's prospects would be bleak.

THIS COMPLETES OUR INVESTIGATION of the cellular you. We have seen how, thanks to cell specialization, your ancestors were able to transform from solitary cellular organisms into the complex, multicellular, macroscopic organism that you are. You consist of 37 trillion cells, but this number is misleading. For one thing, each of what you would regard as your human cells will have, on average, hundreds of "cells"—we call them mitochondria—living inside of them. Furthermore, each of your human cells will have, on average, ten cells—members of your microbiome—living outside of them. Each of these non-human cells has its own DNA, which is different from what you would regard as your DNA. Consequently, a case can be made that the "cellular you" is comprised of not 37 trillion cells, but quadrillions of cells, less than 1 percent of which are "human."

You think of yourself as an organism, but you are in fact an ecosystem. You think of yourself as alive, but you are in fact permeated with living things. And take my word for it: you wouldn't want it any other way.

PART III

The Atomic You

13

You Are What You Eat, Ate

SO FAR, WE HAVE TAKEN YOU TO BE a person, a member of a species, and a collection of cells, but these aren't the only ways in which you can be understood. Ask a physicist what you are, and the likely response will be "a collection of atoms." The physicist might go on to reveal that if you weigh, say, 155 pounds (70 kilograms), you are composed of approximately 6.7×10^{27}—6.7 billion billion billion—of them.[1]

At birth, you might have weighed 9 pounds (4 kilograms) and been comprised of 0.4×10^{27} atoms. Where did the additional 6.3×10^{27} atoms come from? The naive answer to this question, offered by a surprising number of people, is that these atoms "are just there" or that your body somehow made them. Such responses indicate a failure to appreciate one of the most basic principles of physics, the Law of Conservation of Mass: although mass can be moved around, it can't be created or destroyed—not, at any rate, in the normal course of things.[2] This means that you didn't create any of the atoms in you. You simply borrowed them from the outside world. And when you are done with them, that is where they will return.

When an atom enters a living thing, various chemical processes go to work on it. As a result, a solitary atom might become part of a molecule, and an atom that was already part of a molecule might become part of a different molecule. The chemical transformations in question can have spectacular results. Monarch butterflies, for example, are comprised primarily of atoms that formerly belonged to the milkweed leaves that, as caterpillars, they fed on. Likewise, silk is comprised primarily of atoms that formerly belonged to the mulberry leaves that silkworms fed on.

There is an old saying that you are what you eat. Atomically speaking, this is true: the atoms that comprise you almost all come from the foods you eat and beverages you drink.[3] But this aphorism also applies to other animals. Consequently, if you are a carnivore, besides being what you eat, you are what you eat, ate. Eat free-range chicken, and some of the atoms you are ingesting probably belonged to one of the bugs that the chicken hunted down. (Eating such bugs is thought by some to play a role in making free-range chickens tastier than their commercially grown counterparts.) Eat a lobster, and some of the atoms you are ingesting might previously have belonged to the decaying fish that the lobster scavenged.

It is natural to think of the atoms that comprise you as being "your" atoms, but in fact they had long histories before becoming part of you; indeed, some of them have been around for nearly 13.8 billion years. And when you die, these atoms won't cease to exist. They will continue to exist in your remains, or they might move on to become part of some other living thing. In all likelihood they will outlast you by billions of years. Consequently, from your atoms' point of view, your body is simply a way station on a very long journey.

IN ORDER TO GAIN WEIGHT, as I've said, you have to acquire new atoms. This is usually done by eating foods and drinking beverages, but when you inhale, you acquire, if only for a few seconds, some atoms. (A sensitive scale can tell whether your lungs are full of air or empty.) You can also acquire atoms by getting a tattoo, having a cavity filled, or injecting medicines or drugs into your body. Likewise, a bullet that lodges in your flesh will increase your weight by a few grams. And in order to lose weight, you will have to lose some of the atoms that currently comprise you. This can be done by urinating and defecating. It can also be done by exhaling, by crying, by sweating, by vomiting, and by having a decayed tooth or a lodged bullet removed.

Pop quiz: which will cause you to gain more weight, eating a pound of chocolate or drinking a pound of water? Lots of people answer "chocolate," because they know it is a fattening food, but initially the weight gain will be exactly the same: one pound. After that, though, your body will likely lose the water molecules but find places to store the chocolate's fat and carbohydrate molecules. As a result, your weight will probably be higher a week later if you ate a pound of chocolate than if you had instead drunk a pound of water.

Suppose that as the result of eating too much chocolate, you have gained what you take to be an undesirable amount of weight. If you go on a diet, the deposits of fat within you will shrink, but it can't be because atoms of fat are somehow vanishing: atoms, as I've said, don't just vanish. It must instead be that fat atoms are leaving your body. But how? Do you sweat them out or maybe, as some people think, defecate them?

It turns out that when fat is metabolized—when, as we say, it is "burned"—fat molecules are transformed into water molecules and carbon dioxide molecules. The "metabolic water" molecules can subsequently leave in your breath, feces, urine, sweat, or even tears, while the carbon dioxide molecules are exhaled. This means that atomically speaking, it is not only possible to sweat away fat, but to cry or breathe it away. And humans aren't alone in their ability to make water in this manner. Because hibernating bears make metabolic water when they burn their stores of fat, they can get through the winter without drinking. The same is true of birds making long transoceanic flights. Likewise, camels are able to cross deserts without drinking not because they store water in their humps, but because they store fat that can be metabolized into water.

People know that they inhale oxygen and exhale carbon dioxide, but they usually don't go on to ask what should be the obvious question: where do the carbon atoms in that carbon dioxide come from? They could only have come from us, but what part of us? Some of them, as I have said, were in the fats we consumed; others were in the carbohydrates or proteins we consumed.[4] The metabolic process—and here, I am dramatically simplifying things—plucks out these carbon atoms and attaches them to molecules of oxygen[5]—what chemists refer to as O_2—thereby releasing energy. The resulting carbon dioxide (CO_2) molecules enter our bloodstream and are subsequently expelled in our breath. The O_2 molecules you inhale therefore are like porters who enter your body, pick up a load consisting of one carbon atom, and carry it out of your body. With each trip they make, your body loses a carbon atom and thereby becomes lighter.

We obviously burn calories and thereby lose weight when we are exercising, but we also do so when we are at rest—indeed, even when we are asleep. In a typical night of sleep, in which we make no trips to the bathroom or refrigerator, we might lose 8 ounces (227 grams) of weight.[6] Most of this weight loss will be the result of our losing water molecules through the pores of our skin in the form of sweat, or through the lining of our lungs as we breathe, but maybe 2 or 3 of the ounces lost will be due to the carbon atoms we expel when we exhale. Yes, these atoms are individually quite light, but exhale enough of them, and their combined weight is significant.

It is astonishing to think that we might, in the course of a year, lose 180 pounds (82 kilograms) in our sleep. Furthermore, when we are awake, we are likely burning calories and therefore losing atoms at an even faster rate. But of course, when we are awake, we are also likely to be eating and drinking, meaning that most people have no trouble at all gaining back the weight they lost in their sleep—and then some.

HOW MANY POUNDS OF FOOD do you eat in a year? If you answered, flippantly, "A ton!" you probably aren't very far from the truth. According to the United States Department of Agriculture, the average American consumed, in the year 2000, the following foods: 150 pounds of sweeteners, 200 pounds of grain products, 700 pounds of fruit and vegetables, 75 pounds of fats, 600 pounds of dairy products, and 200 pounds of meat. That is 1,925 pounds (873 kilograms) of food—nearly a ton—which works out to about 5.5 pounds (2.5 kilograms) per day.[7]

And to this, we need to add the weight of the beverages consumed annually: on average, 45 gallons of soft drinks, 30 of bottled water, 20 of beer, and 20 of coffee. There might also be 30 gallons of tea, sports drinks, and hard liquor. (Milk consumption was listed above in dairy products.) Call it 150 gallons per year, to which we would add whatever water you consumed from the tap or in the form of ice.[8] This means that you consume maybe 200 gallons of liquids per year—about 1,700 pounds.[9] Bottom line, in the course of a year, you ingest perhaps 3,700 pounds (1,678 kilograms) of food and drink. That's about 10 pounds (4.5 kilograms) a day, of which 5.5 are food and 4.5 are drink.

Many people will find these daily food consumption numbers to be surprisingly high; I know I did. But then I started thinking about how heavy the grocery bags are that a few times a week I lug from the garage to the house. To settle the matter, I started keeping track of my consumption of food and beverages. I kept track not of calories but of the weight of the food I consumed. Furthermore, I weighed my food after it was cooked, and I was careful to subtract out the part of the food I did not eat, such as watermelon rind. I found that over a 24-hour period, I would typically eat and drink 7 or 8 pounds of food and beverages. This suggests that in terms of daily food consumption, I am below average. My body weight, I should add, is also below average.

This means that over the course of my lifetime, I have consumed perhaps 190,000 pounds (86,000 kilograms) of food and drink. Of that, a mere 160 pounds (73 kilograms) of matter remains with me—this is my current weight—meaning that I have retained less than 0.1 percent of the food and drink—and therefore less than 0.1 percent of the atoms—that I have ingested. The other 99.9 percent went back into the world, in the atoms in urine and feces, in sweat and tears, in shed skin cells, in lost, trimmed, and shaved hair, in exhaled water vapor and carbon dioxide, and so forth.

Suppose that at birth, I had been put into a capsule that had within it all the air, water, and food I would need for my lifetime. Suppose that the capsule was sealed, and that it stored and recycled things in a way that prevented any atoms from leaving or entering it. If my weight gain in the capsule matched my gain outside of it, I would, in the course of my lifetime, gain 154 pounds—I weighed 6 pounds at birth—but here is the important thing to realize: the weight of the capsule would have remained the same that whole time. It had to, since no atoms entered or left it.

Here is another way to make this point. If you burn a 30-pound log in a fireplace, there will be very little material left when the fire has burned out—maybe a pound of ash. Where did the other 29 pounds of wood go? Up the chimney, in the form, mostly, of water vapor and carbon dioxide. But if you burned the same log in a hermetically sealed chamber that contained enough oxygen to allow for complete combustion, the chamber—which would contain not only the ashes but the trapped gasses of combustion—would weigh the same as it previously did. To lose weight, something—whether it be a person's body or a piece of wood—must lose atoms.

Most of the atoms you acquire enter your body through your mouth and nose. The molecules in the foods you eat and the beverages you drink are directed down your esophagus, while the molecules in the gasses you breathe are directed down your windpipe and into your lungs. These latter molecules include the nitrogen (N_2) and oxygen (O_2) in the air you breathe. The N_2 molecules you inhale will likely be with you for only a few seconds: some will dissolve into your blood, where they can give you trouble if you are a deep-sea diver, but most will be breathed right back out. The O_2 molecules you inhale, though, will be with you for a somewhat longer period. They will pass through the membranes of your lungs, bind to the hemoglobin in your red blood cells, and be carried to the cells of your body, where they will play the metabolic role I have described.

The solid and liquid molecules that go down your esophagus are much more likely to take up intermediate- or long-term residence within you. When food and beverages go into your stomach, they are mixed with acid and enzymes and churned by muscular contractions until they become the chunky slurry called *chyme*. In this digestive process, carbohydrates are broken into sugars, and fats are broken into fatty acids. Proteins get broken into their component amino acids, which can then be used by your body to build new proteins.

Foods then move from your stomach into your intestines. In your small intestine, nutrients—protein, sugars, fat, salt, etc.—are extracted by little finger-like villi. These nutrients then pass into your bloodstream, which is astonishingly multifunctional. It carries liquids, including all the water needed by your cells, as well as alcohol, if you have been drinking. It carries the amino acids your body needs to make proteins—meaning that it carries the building material not just for muscles, but for hair and fingernails. It also carries the calcium necessary to make bones. It carries the fuel your body needs, in the form of sugars and fatty acids, along with the oxygen necessary to "burn" that fuel. It carries any intravenous medications you might be taking or recreational drugs you might be using. It also carries the waste products of metabolic processes. Some of these are removed by your kidneys and expelled in your urine; others, including primarily carbon dioxide, are expelled by your lungs. Your bloodstream is therefore a combination fuel pipe, water pipe, building material pipe, oxygen pipe, exhaust pipe, and sewage system. How remarkable!

From your small intestine, food moves into your large intestine, the primary function of which is to retain, through absorption, the water you consume. That we need such an organ is an indication that our evolutionary ancestors lived in an environment in which water was somewhat scarce. Fish, by way of contrast, have all the water they need, so they lack large intestines. After the nutrients and water have been removed from what you ate, what is left is feces. It consists of anything that was indigestible, including cellulose. It also contains the remains of microbes that, as we saw in the previous chapter, reside in your gut. Finally, it contains the bile and bilirubin from dead red blood cells. This is why feces is brown.

IF SOMEONE ASKS HOW OLD YOU ARE, you will reflexively tell them how long ago you were born. In my case, as I write this, the answer is 65 years. But as we saw in chapter 9, this is only one of many ages that can be associated with a person. Indeed, in cellular terms, you are (on average) only a child.

So much for your biographical and cellular age. What about your "atomic age"? In answering this question, we need to keep in mind that your atomic age will be determined in a rather different manner than your cellular age. Your cells came into existence within you, almost exclusively as the result of cell fission. Your atoms, however, came into existence billions of years before you were born. When we talk about your atomic age, we are therefore talking about the amount of time, on average, that the atoms that now comprise you have been part of you. And realize that it is a mistake to think that they have been with you since birth. Most of them clearly haven't; otherwise, you would weigh the same as you did when you were born. To the contrary, some atoms have been with you for decades, whereas others joined you in your most recent inhalation.

The obvious way to determine your atomic age would be to "tag" individual atoms to see how long they stay with you. But how, you might wonder, can we tag an atom? One way is to put it into a nuclear reactor or place it in the beam of a particle accelerator. If we do things right, the atom will gain a neutron or two and thereby become radioactive. We can then use a Geiger counter or some other instrument to detect the continued presence of these "tagged" atoms.

By means of this tagging process, scientists have converted ordinary table salt into radioactive salt. Ordinary salt, besides containing chlorine atoms, contains sodium-22 atoms, each of which has 11 protons and 11 neutrons. The radioactive salt instead contains sodium-24 atoms, each of which has 11 protons and 13 neutrons. The extra neutrons in sodium-24 make it radioactive, but because it has the same number of protons as sodium-22, it behaves the same, in chemical reactions, as sodium-22 does. This means that a human body will use radioactive salt in the same way as it uses ordinary salt, which in turn means that by tracing the movement of radioactive salt through a person, scientists can determine how ordinary salt moves.

Tagged-atom experiments became quite popular after World War II. The US government had built reactors as part of its effort to construct the atomic bomb. When the war was over, it embarked on a campaign to show the public that these reactors had peaceful applications as well. They were put to work making radioisotopes that were subsequently made available to researchers who used them in a variety of experiments.

In one of these experiments, subjects closed one hand around a Geiger counter tube that was encased in a lead cylinder and used their other hand to drink a glass of water containing radioactive salt. The salt would pass through their stomach, into their intestines, into their bloodstream, and finally into their tube-holding hand, where its presence would be detected. It took between two-and-a-half and ten minutes for the salt to complete this trip.[10] Researcher Paul C. Aebersold later did a refined version of this experiment in which he injected radioactive saline solution into the arm of a subject. It took only fifteen seconds for it to travel through the subject's heart and lungs and appear in the other arm. In another minute, it had sufficiently diffused into the tissues of that arm to appear in sweat on the surface of the skin.

Aebersold's conclusion: "the atomic turnover in our bodies is quite rapid and quite complete." More specifically, he found that it takes only a week or two for half of our sodium, hydrogen, and phosphorus atoms to be replaced by other atoms of their kind. Our carbon atoms are a bit slower to turn over: in a month or two, half of them are supplanted by other carbon atoms. He goes on to estimate that in the course of a year, "approximately 98 percent of the atoms in us now will be replaced by other atoms that we take in in our air, food, and drink."[11]

It would therefore appear that even though you are old enough to vote—or even old enough to retire—the atomic you is still in its diapers, so to speak. It is also highly unlikely that *any* of the atoms that comprised you when you were a zygote are still part of you, meaning that we are confronted with the ship of Theseus paradox described back in chapter 9: given that the current you and the zygotic you have no atoms in common, are you, atomically speaking, the same person as you were then?

This, as we shall see, is not the last time we will encounter this paradox. The difficulty we have in answering the questions it raises is evidence, perhaps, that although we very much value our personal identity, we aren't quite clear about the nature of that identity.

14

Your Windblown Past

IN THE PREVIOUS CHAPTER, WE SAW that you are, quite literally, what you eat. Almost all of what you call *your* atoms were formerly atoms in the foods and beverages you ingested. This means that we can go very far in telling the story of the atomic you by tracing the histories of the atoms in your meals.

Let us begin with the carbon atoms that comprise 12 percent of you.[1] They entered you in the carbohydrates, fats, and proteins you consumed, which in turn ultimately got them from plants. This is obviously true if you got your carbon atoms from, say, an avocado. If you instead ate a steak, they would have come to you indirectly, from the cow that got them from grass or corn. And where do plants get their carbon atoms? From the air—more precisely, from the CO_2 molecules that make up 0.039 percent of the atmosphere.

These CO_2 molecules in turn could have come from a number of sources. Some came from volcanoes. Others came from animals, including ourselves, that exhaled them after having "burned calories." Besides producing CO_2 molecules in this manner, we humans produce very many more of them in our daily activities. We burn the earth's forests, as well as its fossil fuels, including coal and petroleum. As a consequence of these activities, we have, in the last two centuries, increased the amount of CO_2 in the atmosphere by nearly 40 percent and thereby significantly altered the earth's climate.[2]

In chapter 9, we traced the history of one of your cells, chosen at random. Let us now trace the history of a carbon atom within a randomly chosen carbohydrate molecule within one of your cells. It came to you, as we have seen, in the food you ate. It might have been part of an English muffin that you had for breakfast. Before that, it might have been, going back in time, part of the flour used to make that muffin; part of one of the kernels of wheat used to make that flour; part of a CO_2 molecule the wheat plant took in; part of the prairie dog that, after burning some calories, exhaled that CO_2 molecule; part of the grass that prairie dog ate; part of the CO_2 molecule that grass took in from the atmosphere; part of the tree that, when burned, produced that CO_2 molecule; part of the CO_2 molecule that tree took in; part of the gasoline that, when burned, produced that CO_2 molecule; part of the ancient microscopic plankton that, on dying and being covered with other material, became part of the petroleum that was refined into that gasoline; and part of the CO_2 molecule that microorganism took in. This last molecule might have been belched out by an ancient volcano. This is, to be sure, a fanciful story, but there is every reason to think that the histories of your individual carbon atoms, if they could be traced, would be equally amazing. It is also unlikely that any two of your carbon atoms have the same history.

ALL CARBON ATOMS HAVE SIX PROTONS—that is what makes them carbon—but the number of neutrons can vary. Ordinarily, carbon atoms have six neutrons. Chemists therefore refer to them as carbon-12 atoms (notice that 6 protons + 6 neutrons = 12 nucleons) or, using nuclear notation, as ^{12}C atoms. As we saw in chapter 9, though, it is possible for carbon atoms to carry "extra" neutrons. During the cold war, above-ground testing of nuclear weapons gave rise to carbon atoms that carried two additional neutrons, for a total of 8. It was these ^{14}C atoms (notice that $6 + 8 = 14$) that served as the tracer that let scientists determine the average age of a cell. And even in the absence of nuclear testing, about 1 percent of the earth's carbon atoms carry one additional neutron, for a total of 7. The existence of these ^{13}C atoms (notice that $6 + 7 = 13$) enabled scientists to explore the rather surprising role that corn plays in your diet. It turns out that there are different forms of photosynthesis. Most plants do what is called *C3 photosynthesis*, but many grass plants, including sugarcane, sorghum, and corn, have gained the ability to do the rather more efficient form of photosynthesis known as *C4*. These C4 plants, when they take in CO_2 molecules, favor those that contain the ^{13}C isotope of carbon over those that contain ^{12}C. This means that the more corn you eat, the higher the ratio of ^{13}C to ^{12}C atoms within you will be. (The sorghum and sugarcane-derived sugar you eat will also be sources of ^{13}C atoms, but these foods are unlikely to play as significant a role in your diet as corn does.) This in turn means that by examining the ratio of ^{13}C to ^{12}C atoms in you, scientists can estimate how much corn you consume.

News anchor Diane Sawyer, in conjunction with a story, had her carbon ratio tested. It turned out that 50 percent of her carbon atoms came from corn.[3] You might think that you eat very little corn—that maybe Sawyer is a corn-on-the-cob or popcorn fanatic. But realize that besides the "obvious" ways to ingest corn-derived carbon atoms, there are other, inconspicuous ways. Suppose you order a taco and soft drink at a Mexican restaurant. You will obviously be ingesting the corn-derived carbon atoms in the corn in the taco shell. Less obviously, you will be ingesting the corn-derived carbon atoms in the high-fructose corn syrup that the soft drink likely contains. And not to be forgotten, you will be ingesting the corn-derived carbon atoms in the hamburger that the taco is filled with. Corn, after all, is a principal ingredient in animal feed: one study found that 93 percent of the carbon in the hamburger it sampled had come from corn.[4] Consequently, people who eat lots of beef indirectly eat lots of corn and are therefore likely to end up with an excess of ^{13}C atoms, meaning that it is entirely possible, and deliciously ironic, that a carnivore can end up being "cornier" than a vegan.

SO MUCH FOR YOUR CARBON ATOMS. What about the hydrogen and oxygen atoms that comprise, respectively, 62 and 24 percent of you? In answering this question, it is useful to distinguish between the hydrogen and oxygen atoms that entered you as components of the water molecules you consumed and those that entered you as components of the fat, carbohydrate, and protein molecules you consumed. These last atoms can ultimately be traced back to plants, and we will turn our attention to them in a moment, but first, let us explore the history of the hydrogen and oxygen atoms that entered you as components of water molecules.

You are obviously consuming water if you drink a glass of the stuff. There is also water in almost any beverage you drink—a glass of pure alcohol would be an exception. And even if you eat rather than drink, you are almost certainly consuming water. Cucumbers, for example, are 95 percent water, meaning that you could stay alive without ever drinking anything, as long as you had enough cucumbers to eat. Meat is also mostly water, which is why one pound of beef might dehydrate into only one-third pound of beef jerky. There are foods that don't contain water—sugar cubes, for example—but man does not live by sugar cubes alone.

Because water molecules are quite stable, they can endure for millions or even billions of years. This means that when you drink a glass of water, it is possible that some of its molecules have been drunk before. It is possible, for example, that a bit more than four centuries ago, one of the molecules in your glass was part of some ale quaffed by William Shakespeare, who subsequently put it back into circulation by shedding a tear. Go back 67 million years, and it is possible that this very molecule was part of a gulp of water swallowed by a *Tyrannosaurus rex,* which put it back into circulation by urinating.

The water molecules you are drinking have likely been in oceans and clouds. They have fallen as components of raindrops and snowflakes. They have spent time in rivers and lakes, and some of them might have spent thousands of years in glaciers or millions of years deep underground. It is also possible, though, that some of the water molecules in the glass you are drinking came into existence only a few days before, in the engine of a car. Allow me to explain.

Gasoline is a mixture of hydrocarbon molecules. In the engine of a gasoline-powered vehicle, the carbon atoms of these molecules combine with O_2 molecules from the car's air intake to produce CO_2 molecules,[5] and the hydrogen atoms of the hydrocarbon molecules combine with oxygen to produce water (H_2O) molecules. These combustion products are vented from the car's exhaust system. When released on a cold winter morning, the water molecules stick together to form water droplets, and as a result, we see "smoke"—more properly characterized as fog—coming out of cars' exhaust pipes. It soon disappears, as the water droplets evaporate to become invisible molecules of water vapor. On warm days, the water molecules aren't similarly sticky: they don't condense to form droplets, and cars are consequently "smokeless."

It is therefore possible that one of the molecules in a glass of water that you drink came into existence in the engine of a car. It passed into the atmosphere, became part of a cloud, fell to earth in a raindrop, and was sucked from a lake by your local water company, which piped it to your house, where you poured it into your glass. This in turn means that when you drink this molecule, you will be ingesting two hydrogen atoms that not long before were components of a hydrocarbon molecule in the gas tank of a car. And millions of years before that, the atoms in question were components of microscopic plankton that, after their death, fell to the ocean floor where they were buried and slowly transformed into the petroleum from which the gasoline was refined. Atoms, as we shall see in the next chapter, have very long histories.

SO MUCH FOR THE HYDROGEN AND OXYGEN ATOMS that entered you as components of water molecules. What about those that entered you, instead, as components of fat, carbohydrate, and protein molecules in the foods that you consumed? Even if you are a carnivore, these atoms, as we have seen, can be traced back to plants. This means that by looking into how plants get their hydrogen and oxygen atoms, we can gain important insight into the history of your own hydrogen and oxygen atoms.

We have seen that plants get their carbon atoms from carbon dioxide molecules, but what is the source of their hydrogen and oxygen atoms? They get them by performing photosynthesis: they take molecules of carbon dioxide and water as inputs, and chemically combine them in the presence of light to produce molecules of glucose and oxygen. The glucose is subsequently used to form other carbohydrates, as well as fats and proteins. The process can be summarized by this chemical equation:

$$CO_2 + H_2O + \text{light energy} \rightarrow C_6H_{12}O_6 + O_2$$

Although this is the equation that many of us learned—or at least were exposed to—in biology class, the chemistry of photosynthesis turns out to be rather more complicated than the equation implies.

For one thing, besides producing O_2, plants consume it when they carry out their own metabolic activities; and besides consuming CO_2, plants produce it, again as a result of their metabolic activities. In the daytime, if it is sunny, a plant will be a net consumer of CO_2 and net producer of O_2, but at night, these roles will reverse.

Fortunately for us, because of their sedentary lifestyle—plants don't move around the way we animals do—plants produce a lot more oxygen than they consume. Otherwise, our atmosphere wouldn't have as much oxygen as it does, and we animals would be in a bind.

For another thing, engaging in photosynthesis requires a lot more than simply bringing together carbon dioxide and water molecules in the presence of sunlight. You would find this out if you put some carbon dioxide and water into a bottle, sealed it, and then set it out in the sun alongside your tomato plants. At the end of the growing season, your plants would have brought forth dozens of delicious tomatoes, but the bottle would still contain water and carbon dioxide. Photosynthesis, it turns out, requires multiple chemical steps and involves some dauntingly complex chemistry.[6]

And finally, for a plant to carry out photosynthesis, it isn't enough to provide it with *some* CO_2 and H_2O molecules. It needs precisely 6 of each in order to make a single molecule of glucose. A more accurate way to describe this process is therefore with the following chemical equation:

$$6CO_2 + 6H_2O + \text{light energy} \rightarrow C_6H_{12}O_6 + 6O_2$$

Notice that this equation is balanced, inasmuch as it has 6 carbon atoms, 12 hydrogen atoms, and 18 oxygen atoms on both its left and right side.

Not even this equation, however, gets to the bottom of things, chemically speaking. A more accurate way to describe the photosynthetic process is as follows:

$$6CO_2 + 12H_2O + \text{light energy} \rightarrow C_6H_{12}O_6 + 6O_2 + 6H_2O$$

This formula initially looks redundant. It shows 12 water molecules as input and 6 as output. Why not just show 6 molecules of water as input and none as output, as the previous formula does? Because if you started out with fewer than 12 molecules of water, the process of photosynthesis would grind to a halt; all 12 come into play at various points in the reaction. Consequently, to list all 12 as reactants is not redundant; it is instead an accurate representation of the chemical process that takes place.

Furthermore—and this is quite significant in telling your atomic history—the 6 water molecules that are the result of the photosynthetic process will individually *be different from any of the 12 water molecules that the process started with.* This is because the 6 final H_2O molecules will have gotten their hydrogen atoms from the original 12 H_2O molecules and their oxygen atoms from the original 6 CO_2 molecules (see Figure 14.1). Each of the final 6 H_2O molecules will therefore have different oxygen atoms than any of the first 12 H_2O molecules, meaning that *they will be different molecules.* The glucose molecule, by the way, will have gotten its hydrogen atoms from the original 12 water molecules and its carbon and oxygen atoms from the original 6 carbon dioxide molecules. Scientists figured these things out by the clever use of tagged atoms.[7]

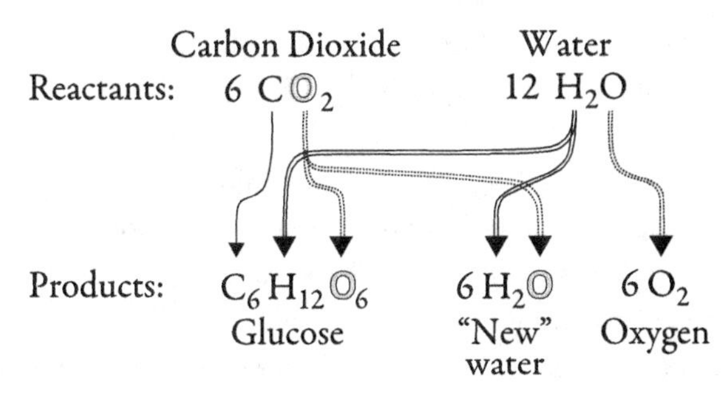

FIGURE 14.1. This is the chemical reaction that takes place in photosynthesis. The surprising thing is that all the oxygen released into the atmosphere comes from the reactant water molecules; none comes from the carbon dioxide. The oxygen in the carbon dioxide—I highlight this oxygen by using a different font—all goes into the glucose and water molecules that are manufactured. This means that the water molecules that are produced by photosynthesis are composed of different oxygen atoms than the water molecules that served as raw materials for the photosynthetic process, meaning that they are "new" water molecules. Finally, the hydrogen in the original water molecules ends up both in the glucose molecule and in the "new" water molecules.

And while we are discussing the identity of water molecules, a confession is in order. Earlier in this chapter, I said that water molecules are durable things that can last for billions of years. In doing so, I was glossing over some technicalities. It is true that if you isolate a water molecule from other water molecules, it can last for billions of years, but put it into contact with other water molecules to make liquid water, and something amazing happens: the molecules start swapping hydrogen atoms with their neighbors. And such swaps are by no means occasional things. They might happen to a water molecule 10^{11} times in a single second.[8] This means that if you think, quite sensibly, that the identity of a molecule is determined by its component atoms, you will find it very difficult to trace the history of your water molecules. You will constantly be confronted with molecular identity crises that are, in essence, the ship of Theseus paradox writ exceedingly small.

WE ANIMALS CLEARLY BENEFIT from the existence of plants. We breathe in the oxygen and eat the food they produce. They in turn take in the carbon dioxide we exhale as a waste product and chemically combine it with water to make more oxygen and food for us. Indeed, if it weren't for plants, we animals wouldn't exist. There would be no oxygen for us to breathe, and even if we had oxygen, what would we eat?

Of course, at the same time as we are benefitting from plants, they are benefitting from us. They get much of the CO_2 they need from animals in general and in particular from human beings: we provide them not only with the CO_2 we exhale but with the CO_2 from all the fossil fuels we burn. And in many cases, we also provide them with the H_2O they need to carry on photosynthesis—in other words, we water them.

Some plant species have not just benefitted but have positively thrived because of their relationship with humans. Without us, for example, corn plants would be pathetic little weeds scattered around the landscape. Thanks to us, they have been genetically transformed into giants, and because we find corn so useful, we have cleared forests to make fields in which it can grow. We fertilize that corn. We protect it against competition, in the form of weeds, and against predators, in the form of insects. Recently, the world has grown enough corn annually to provide every man, woman, and child on Earth with more than 300 pounds (136 kg) of it.[9] That's a lot of corn!

Sometimes, when I am driving across America's heartland in July and see corn growing in rows that extend to the horizon, I get a sneaking suspicion that we humans are being taken advantage of. We work so hard for the benefit of plants! This suspicion is reinforced when I read news reports about marijuana growers being busted. We humans find marijuana plants sufficiently useful that we are, in some places, willing to run the risk of a prison sentence to construct grow rooms in which we water them, feed them, provide them with the ideal amount of light, and even adjust the amount of carbon dioxide in the air around them. With respect to marijuana, we play the role of lackeys—or maybe patsies would be the more appropriate word.

WE HAVE NOW EXPLORED THE SOURCE of your carbon, hydrogen, and oxygen atoms. What about the nitrogen atoms that comprise 3 percent of you, though? They are a component of the amino acids that make up your proteins. Without nitrogen atoms, you would not have muscles, hair, or fingernails; indeed, you would be dead.

You get your nitrogen atoms the same place as you get your carbon atoms—namely, from plants, either directly, by eating plants themselves or indirectly, by eating animals that ate plants. It is tempting to think that plants would get their nitrogen the way they get their carbon, by "inhaling" it—the atmosphere is, after all, composed mostly of N_2 molecules—but this turns out not to be the case. Because N_2 molecules are chemically inert, plants can't make use of the nitrogen they contain.

What plants do instead is get nitrogen atoms through their roots, in the form of nitrates that are dissolved in the water they draw up. These nitrates come from the air, but in an exotically indirect manner—actually, in one of three indirect manners. The first involves nitrogen-fixing bacteria, which have the ability to convert airborne N_2 molecules into ammonia. The roots of some plants, especially legumes, have nodules that contain these bacteria. These plants provide the resident bacteria with a nice place to live, and in return, the bacteria provide them with the nitrogen compounds they need to grow. Lightning is a second source of nitrates. When a lightning bolt rips through the atmosphere, it transforms the N_2 atoms it encounters into nitrates, which are dissolved into raindrops that subsequently fall to the ground, where they and the nitrogen they contain are taken in by plants. The third means by which atmospheric nitrogen is "fixed" involves the ingenuity and efforts of humans. Early in the twentieth century, chemist Fritz Haber invented the eponymous process that puts air (and the nitrogen it contains) and natural gas into a closed reactor, where they are heated and put under great pressure. As a result, nitrates are formed. This process is the primary source of the inorganic fertilizer used by farmers and gardeners.

By eating a juicy steak for dinner, you will acquire water molecules and carbon atoms. You will also acquire the nitrogen atoms that are components of the steak's protein molecules. Before one of the steak's nitrogen atoms became part of you, it was part of a cow. Before that, it might have been part of a blade of grass; before that, part of a nitrate molecule in the soil; and before that, part of an N_2 molecule in the air. But between being in the air and being in the plant, that nitrogen atom likely experienced one of three fates: it was taken in by bacteria living in nodules in the roots of a plant, it was struck by lightning, or it was put through the Haber process in a fertilizer factory. Like I say, if your atoms could talk, they would have amazing stories to tell.

IF YOU TOOK A CENSUS OF YOUR ATOMS, you would find that hydrogen, oxygen, carbon, and nitrogen atoms constitute 99 percent of you. This means that by telling how your hydrogen, oxygen, carbon, and nitrogen atoms came to you, we have gone far in providing a short-term atomic history of you. To be sure, we have not provided a detailed and individualized history of your atoms—that would be impossible to do. We have nevertheless gained many important insights into the past whereabouts of the atoms you call yours.

Significantly, the four most common elements in you are among the six most common elements in the universe. It probably isn't an accident that you are composed primarily of readily available elements. It also isn't an accident that the other two top-six elements in the universe—namely, helium and neon—aren't part of you. They are "noble gasses": because of how their electrons are configured, they are chemically inert, meaning that they are reluctant to combine with other atoms to make molecules. This makes them of limited use in life processes.

You likely think of yourself as a solid entity, but since you are, by weight, mostly water, it is more accurate to describe you as a liquid being. And if we explore your atomic history, we might instead describe you as a windborne being. Your water molecules were formerly airborne molecules of water vapor, meaning that their hydrogen and oxygen atoms were airborne. Likewise, the carbon, hydrogen, and oxygen atoms in your carbohydrate, fat, and protein molecules were formerly components of airborne carbon dioxide and water vapor molecules, and the nitrogen atoms in your protein molecules were formerly components of airborne N_2 molecules. Conclusion: not so long ago, the carbon, hydrogen, oxygen, and nitrogen atoms that make up 99 percent of you were blowing in the wind.

This brings us to the last 1 percent of your atoms. They include atoms of the 20 or so other elements that are essential for life, including sodium, chlorine, calcium, and potassium.[10] Your sodium and chlorine atoms probably entered you as components of salt, and the salt in question probably came from the ocean. (Realize that even if you got your salt from a salt mine, that mine would likely have gotten its salt long ago, from an evaporating ocean.) Your calcium atoms probably came from the dairy products you consumed, which in turn came from cows. And whether these cows got their calcium atoms from the grass they ate, the corn they were fed, or the nutritional supplements dairy farmers gave them, the atoms in question ultimately came from the ground. Your potassium atoms likewise came from foods you have eaten, such as potassium-rich potatoes and bananas, which also got them from the ground.

But of course, we once again haven't gotten to the bottom of things. Yes, you got your sodium, chlorine, calcium, and potassium atoms from the earth's crust and oceans, but where did the earth get them? Let us, at this point, cut to the chase and ask not particular where-questions, but the most fundamental where-question possible: where did everything, meaning all the atoms in the universe, come from? Again, a complete answer to this question is not possible, but as we shall see in the next chapter, a partial but nevertheless remarkable answer can be given.

15

Your Cosmic Connection

IN THE BEGINNING, THERE WAS NOTHING—no matter, and significantly, no space for it to occupy. There was probably no time either, meaning that it is pointless to talk about what happened *before* the Big Bang event that gave rise to our universe.[1]

Within the first 10^{-32} of a second of the Big Bang, space was created. In the second after that, protons and electrons had formed, some of which subsequently merged to form neutrons. A few minutes later, protons and neutrons started combining to form complex atomic nuclei.[2] A proton and neutron might fuse to form a ^{2}H nucleus, also known as deuterium. (Be careful not to confuse ^{2}H, which is a single hydrogen atom that has an added neutron, with H_2, which is a molecule comprised of two hydrogen atoms.) Two of these deuterium nuclei might then fuse and eject a neutron to form a ^{3}He nucleus, the isotope of helium that has two protons and one neutron; alternatively, they might fuse and eject a proton to form a ^{3}H nucleus, the isotope of hydrogen that has one proton and two neutrons. After that, a ^{2}H nucleus and ^{3}He nucleus might combine and then eject a proton, leaving a nucleus with two protons and two neutrons—leaving, in other words, ^{4}He, the normal form of helium. How, exactly, nuclei can combine and what happens to them thereafter is determined by the rather complex laws of particle physics.

Twenty minutes into its existence, then, the universe contained just two elements, hydrogen and helium. The hydrogen was sorted into three different isotopes: ^{1}H, ^{2}H, and ^{3}H, with one proton but with zero, one, and two neutrons, respectively. The helium was sorted into two isotopes: ^{3}He and ^{4}He, with two protons but with one and two neutrons, respectively. These atoms were spread through space as a gas, and they would have lacked electrons. It was simply too hot for electrons to stay put. Indeed, it took 380,000 years for things to cool enough that electrons could combine with nuclei to produce electrically neutral atoms, in the event misleadingly called *recombination*.[3]

ONE MIGHT THINK THAT having gotten off to such a good start, the fusion process would continue, with the aforementioned nuclei combining to make ever-heavier elements. And it did continue, in a limited manner: it also produced a smidgen of lithium, with three protons, and even less beryllium, with four. At this point, though, the fusion process encountered a major obstacle: when the existing atoms fused, the resulting atoms were unstable and soon broke apart.[4] As a result, for millions of years after the Big Bang, the universe's matter consisted almost exclusively of thinly spread hydrogen and helium atoms. What a boring place!

Obviously this obstacle was overcome; otherwise, we humans, whose bodies are comprised primarily of oxygen and carbon, wouldn't be here to talk about it. The creation of heavier elements required the formation of stars, which happen to be much better vessels for the fusion of atomic nuclei than the early universe was.

For stars to form, the gas of the early universe had to contract to form clouds, and gravity is the only force that could make them do this. Because atoms have so little mass, though, the gravitational attraction between them is minimal. Furthermore, if gas is distributed evenly, atoms won't feel strongly drawn in any particular direction. Indeed, if the matter in the universe had been distributed in a *perfectly* uniform manner, each atom would feel equally attracted in all directions, and as a result, it wouldn't move at all.

Fortunately for us, there were irregularities in the distribution of atoms in the early universe. It isn't clear why these irregularities would have existed—they were the result, perhaps, of quantum effects—but because they did exist, there were atoms that felt an asymmetrical tug from their neighboring atoms. As a result, they drifted, thereby making the distribution of atoms even more irregular. This triggered a snowball phenomenon: the denser a region got, the more it would attract material, meaning that it became denser still.

As a cloud contracted under the influence of gravity, its gas heated up and as a result resisted further contraction. For gravity to win this battle, the compressed gas had to cool—it had, in other words, to radiate its heat. Such radiation is easy if a gas is composed of molecules. Because molecules consist of multiple atoms, they have a two- or three-dimensional structure that allows them to vibrate in a number of ways that a solitary atom can't. Consider, for example, a water molecule. It is shaped like a broad letter V, with the oxygen atom at the vertex of the V and the two hydrogen atoms at the ends of its arms. These arms can vibrate in and out. The angle of the V can also vibrate, becoming bigger and smaller. When a water molecule vibrates in this manner, it radiates heat, in the form of infrared radiation.

Molecules were scarce in the early universe, though. The helium atoms that were present resisted linking up with any other atoms, and the hydrogen atoms were capable of forming only very simple molecules, with H_2, consisting of two hydrogen atoms, being the most common. This meant that gas clouds were slow to cool, meaning that stars were slow to form in the early universe. As a result, for tens or maybe even hundreds of millions of years after the Big Bang, the universe experienced a protracted dark age during which there was plenty of gas but no stars.

If you could travel back to this time, though, you would not have witnessed complete darkness, since the gas around you would have been quite hot. You would instead have seen a reddish glow in any direction you looked. This glow, by the way, is still there, but because the universe has cooled substantially since the Big Bang, it is no longer a reddish glow or even an invisible-to-the-naked-eye infrared glow. Its wavelength has gotten so long that it is in the microwave portion of the electromagnetic spectrum. This "glow" therefore goes by the name of *cosmic microwave background radiation*.

The Cosmic Dark Age, as it might be called, came to an end with the ignition of the first star. It would have been a spectacular sight: one brilliant point of light in an otherwise dully glowing universe. And before long, that star had companions.

WHAT A COLLAPSING CLOUD OF GAS becomes depends on its size. A small cloud might form a gas planet. A big cloud of gas, though, will collapse to form a star. In its core, there will be incredible pressure, and as a result atomic nuclei will be forced to fuse. What kind of fusion reaction takes place depends on the size of a star. In our sun, the primary reaction is the proton-proton chain reaction: four protons (hydrogen nuclei) fuse to make one helium nucleus. In the process of this happening, two of the protons are transformed into neutrons. In stars heavier than the sun, other fusion reactions are also possible. In the triple-alpha process, three helium nuclei fuse to make one carbon nucleus, and in the carbon-nitrogen-oxygen cycle, carbon atoms are transformed into nitrogen by the addition of a proton. The nitrogen atoms are then transformed into oxygen by the addition of another proton.

If a star is bigger still, the fusion process will continue, thereby creating elements heavier than oxygen. In these reactions, energy will be released. But fusing atoms heavier than iron doesn't release energy; it consumes it. This means that once a star's core becomes rich in iron, the fusion process grinds to a halt.

We have seen that gravity draws a star's matter inward and that thermal energy pushes it outward. When a star's fusion process stops, though, its source of thermal energy is cut off, meaning that gravity wins the battle. What happens next depends on the size of the star. If it is the size of our sun, it becomes a red giant. You have seen such a star if you have looked at the constellation of Orion in the night sky. That star is Betelgeuse, the red star at Orion's "shoulder." Being a red giant is only a transitional phase, though, since the star will subsequently blow off its outer envelope of gas to become what is called a *white dwarf*. This, in about 5 billion years, will be the fate of the sun.

If a star is substantially bigger than our sun, it will explode—except that the word *explode* does not do justice to the event that takes place. A single star might, in an instant, release more energy than all the stars in its surrounding galaxy. It is this explosion, known as a *supernova*, that liberates the heavier elements that were produced in the star's core.

In the previous chapter, we talked about your windblown past. We saw that the hydrogen, carbon, oxygen, and nitrogen atoms that make up 99 percent of you were at one time airborne. But the preceding discussion makes it clear that this wasn't the first time your carbon, oxygen, and nitrogen atoms[5] were blowing in the wind. They were also blowing in the stellar winds created by the explosion of the star that forged them. To call these winds ferocious would be an understatement: whereas a hurricane can move air at 150 miles (240 kilometers) per hour, a supernova might move a star's contents at 20,000 miles (32,200 kilometers) per *second*. And yet your atoms survived this experience in order, very much later, to become part of you.

WE HAVE NOW REACHED THE POINT at which the universe contained elements heavier than helium. We have seen, though, that stars cannot, by means of their ongoing fusion process, make elements heavier than iron. How, then, are we to explain the existence in our universe of such elements, including the gold in the fillings of your teeth and the iodine that your thyroid must have, in trace amounts, for you to avoid developing a goiter?

One theory is that these heavier elements are created during supernova events. The collapse of the core of a star releases a flood of protons and neutrons that, when they encounter the nuclei of elements, transforms them into heavier elements. Another, more recent theory is that the heavier elements, and in particular gold, were produced in collisions between what are known as neutron stars.[6] These stellar corpses are left behind when a star four to eight times bigger than the sun explodes in a supernova event. The outer layers of the star are blown into space while the atoms in its core are put under so much pressure that their protons and electrons combine to make neutrons. The resulting matter consists of very tightly packed neutrons and is therefore incredibly dense: one teaspoon of neutron star matter would weigh 900 times more than the pyramid at Giza! Your carbon, oxygen, and nitrogen atoms may have gone through a lot in order to join you, but the gold atoms in your fillings and the iodine atoms in your thyroid have a story that easily tops theirs.

The matter liberated by a supernova is scattered into nearby space to become a stellar debris field. We have one of them in our cosmic neighborhood, a mere 6,500 light-years away. (To put this distance into perspective, the Milky Way is 100,000 light years in diameter.) Astronomers call it the *Crab Nebula* (see Figure 15.1). The supernova that produced this debris took place 7,500 years ago. Its light reached us in 1054 AD and was noted by Chinese astronomers. Modern observations have revealed the existence of a neutron star in the middle of the Crab Nebula. Although this star has more mass than the sun, that mass is packed into a sphere maybe 10 miles (16 kilometers) in diameter.

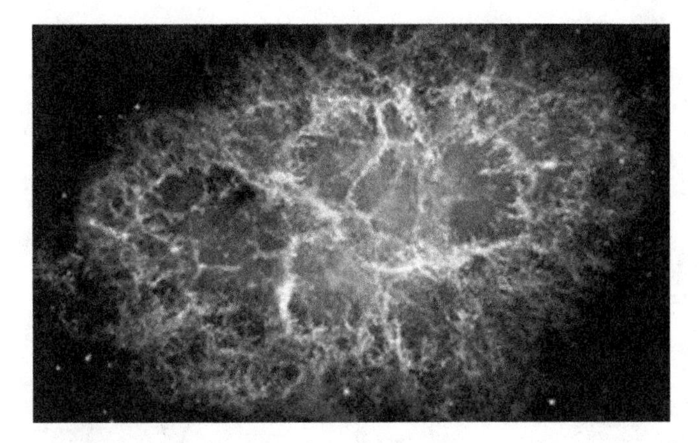

FIGURE 15.1. The Crab Nebula, the debris field of a supernova.

When the universe's first generation of stars[7] blew themselves to bits, the resulting material was mixed with hydrogen and helium atoms that were virgin, in the sense that they had never been components of a star, to provide the material for second-generation stars. As we have seen, the clouds that formed the first stars had few molecules. This meant that they had a hard time radiating heat so they could collapse to form stars. The clouds containing the debris of these stars, however, contained heavier elements that allowed for the formation of many complex molecules, the vibrations of which allowed the clouds to radiate heat effectively and therefore collapse much faster than the first clouds did. As a result, it became much easier for stars to form, and the stars that did form tended to be much smaller than the first-generation stars. They also tended to have much longer lifespans than those stars.

HUMAN FAMILY TREES, AS WE HAVE SEEN, are relentlessly binary. Humans necessarily come from two people, a mother and father, who in turn came from two people. Stellar "family trees," which show the ancestry of stars, are more variable. As our discussion of first-generation stars demonstrates, it is possible for a star to have no stellar ancestors at all. Such a star will be formed out of hydrogen and helium that have never been part of a star. It is also possible for a star to have only one "parent," in the sense that it formed from the supernova debris of only one star. This will be true of many second-generation stars. Most stars will have more than one parent, though. This will be the case if a star formed from the material in overlapping supernova debris fields. One can even imagine a star composed of the debris of dozens of supernovas. In such a case, we might, in the stellar family tree we construct, talk not about parenthood but the percentage of parenthood, with one star being a 21 percent parent, another being a 3 percent parent, and so on. Furthermore, one big star—as the earliest stars would likely have been—can have multiple offspring: it can create enough debris for the formation of dozens of stars.

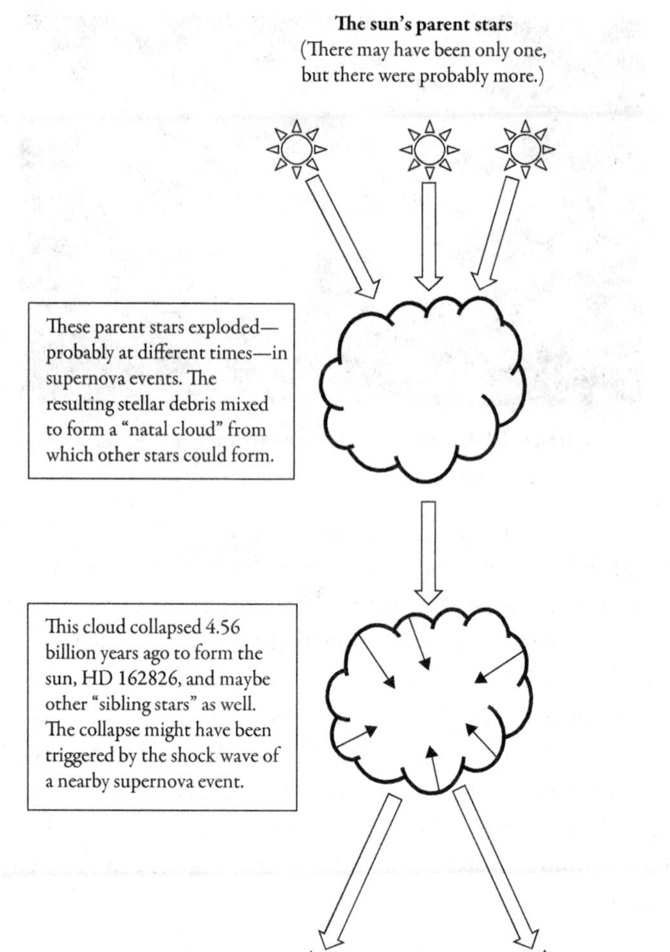

FIGURE 15.2. Astronomers have a rough idea of the structure of the sun's "family tree." They can tell from the sun's chemical composition that it had at least one "parent star." They have also determined that the sun has at least one sibling star, formed out of the same debris field as the sun was.

To determine whether two people are related, genealogists make use of DNA evidence: the more DNA two people have in common, the more closely they are related. Stars don't have DNA, of course, but they do have distinctive chemical makeups. What elements a star contains, and in what relative amounts, will depend on which exploded stars contributed material to its "natal cloud." If two stars have different chemical makeups, they probably formed in different debris fields and therefore have different parents. Conversely, if two stars have very similar chemical makeups, it is likely that the matter from which they formed came from the same debris field, meaning that they are stellar siblings.

Fortunately for astronomers, it isn't necessary to visit a star to determine its chemical makeup. They can do so from a distance by doing a spectral analysis of the star's light. Such analyses indicate that the sun does not have the chemical makeup one would expect in a "parentless" star. Conclusion: the sun had one or more parent stars. Astronomers have also found a star—they call it *HD 162826*—that has a chemical composition very much like that of the sun. This fact, together with the star's location, suggests that it was formed from the same supernova debris as the sun, making it in some sense the sun's "brother star."[8] And taking this line of thought one step further, in much the same way as we construct family trees for people and even for the cells that comprise them, we can construct "stellar family trees," showing a star's ancestors, descendants, and siblings. The sun's tree would look something like the diagram in Figure 15.2.

After their birth, the sun and its twin took different paths through the universe. As a result, the twin is now 110 light-years away. This might seem like a very great separation, but we need to remember that the sun's twin has had a very long time to travel—more than 4.56 billion years, the age of the solar system. Furthermore, for it to end up 110 light-years away, it had to drift at an average speed of only 16 miles (26 kilometers) per hour. This, I should add, is a cautionary tale for siblings everywhere: a very slow drifting apart, continued for a long enough time, will result in a separation that makes reunification impossible.

16

Pulling Yourself Together

WHEN LAST SEEN, in the previous chapter, your atoms were solitary things, floating in the debris field of a supernova explosion. Your hydrogen atoms would have been formed in the Big Bang event, and your heavier atoms, including the carbon, oxygen, and nitrogen atoms that make up so much of you, would have been formed in one or more stars and liberated when those stars blew up.

Your atoms would have found themselves drawn by gravity into a cloud that might have been several light years across and would have about the same elemental composition as the solar system now does: roughly 71 percent hydrogen, 27 percent helium, and 2 percent heavier elements.[1] Because that cloud had a rotational component, it flattened into a disk as it collapsed. The sun, which accounts for 99.85 percent of the mass of the solar system, subsequently formed in the center of that disk, and the planets formed in the disk itself. This rotating-disk theory explains why the planets revolve around the sun in the same direction—counterclockwise, as seen from above the earth's North Pole—and in approximately the same plane. It also explains why the sun rotates counterclockwise on its axis, as do almost all the planets.

Although the atoms in the debris field of a supernova would initially have been solitary, they would subsequently have combined to form molecules, which then combined to form dust grains. These grains in turn might have been drawn together by static electricity and on colliding might have stuck together. Astronauts observed this phenomenon while aboard the space shuttle.[2] As part of an experiment to see what salt grains do in zero gravity, they put some into an inflated plastic bag. Instead of floating independently, the grains clumped together to form what looked like dust bunnies (see Figure 16.1). There

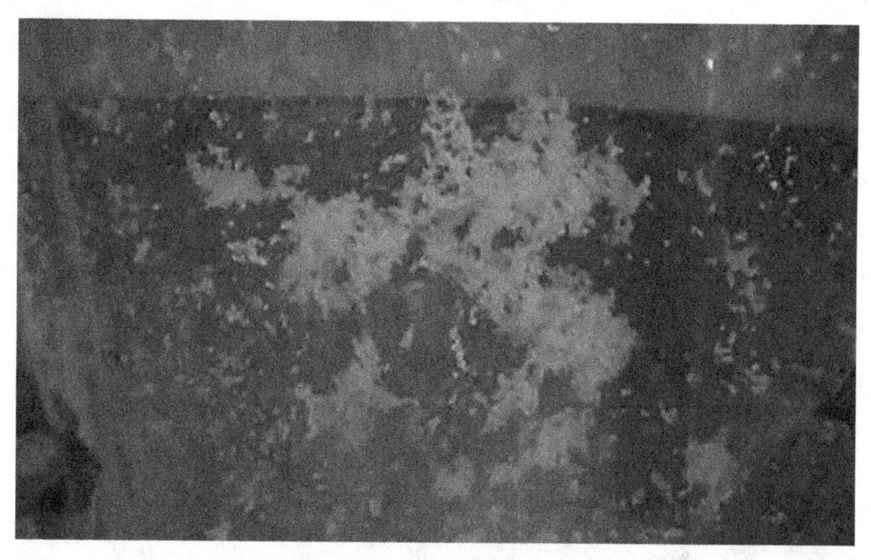

FIGURE 16.1. The "dust bunnies" made by salt crystals floating in an inflated plastic bag at zero gravity. An astronaut's thumb is visible at the lower left.

is reason to think that the atoms that now comprise you were at one time part of similar dust bunnies.

The inner part of the planetary disk was warm enough for hydrogen compounds, such as water and methane, to exist in a gaseous state. As solitary molecules, they would have been susceptible to the solar wind and would therefore have been pushed away from the sun. Farther out, though, temperatures were low enough for these gasses to condense, freeze, and aggregate. This is why the inner planets of the solar system—Mercury, Venus, Earth, and Mars—are composed mostly of heavier elements, while the outer planets are composed mostly of hydrogen and helium.

Once they had formed, cosmic dust bunnies would have been fairly cohesive bodies. If two collided, they were likely to combine and become more compact rather than breaking apart. The bigger they got, the bigger a target they became for the accretion of matter. With the passage of time, cosmic dust bunnies would have grown into cosmic clumps, then cosmic rocks, then cosmic boulders, and so on. As they got bigger, their gravity got stronger, which in turn accelerated their growth into planetesimals.

It is unlikely, by the way, that these larger bodies would have been spherical. They might instead have been shaped like a jagged potato, the way many asteroids are. It was only when a body had grown to a few hundred miles in diameter, as the asteroid Ceres did, that it would have enough gravity to systematically move material from higher to lower elevations and thereby shape itself into a spheroid (see Figure 16.2).

When it first formed, the earth would have been molten: it accumulated its matter by being bombarded by asteroids, and their kinetic energy would have been converted

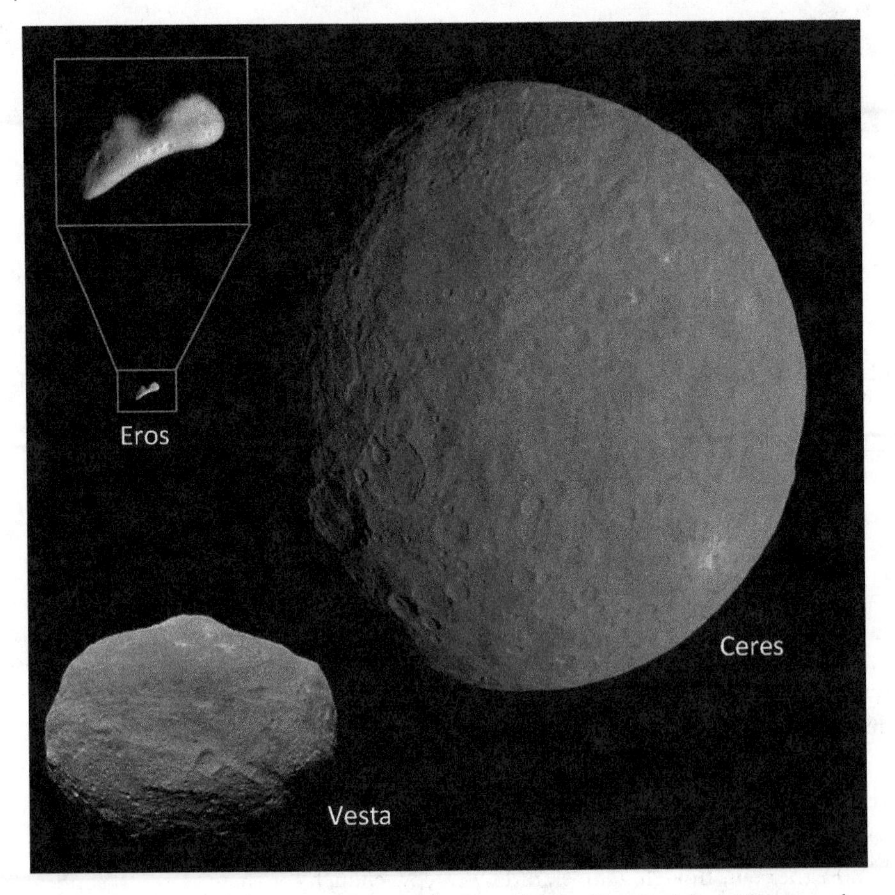

FIGURE 16.2. A size and shape comparison of the asteroids Eros, Vesta, and Ceres. The general principle: the bigger, the rounder.

into thermal energy. When the bombardment slackened and the planet cooled, a crust formed, and the first rocks appeared. Although the solar system is 4.56 billion years old, the oldest known terrestrial rock is "only" 4.374 billion years old.[3]

To better understand the process by which geologists determine the age of rocks, consider those that contain zircon crystals. The molecules of these crystals are normally comprised of atoms of zirconium, silicon, and oxygen. When the crystals form, though, uranium atoms, if they are present, can take the place of zirconium atoms. Once they are embedded in a crystal, these uranium atoms undergo radioactive decay and turn, very slowly, into lead atoms. Since scientists know the rate at which uranium atoms decay, they can, by measuring the ratio between uranium and lead atoms in a zircon crystal, determine how long ago it formed[4] and thereby give us an indication of when the rock containing that crystal formed—when it, in other words, solidified.

At this point, some clarification is in order. In saying that a rock is 4.374 billion years old, geologists are *not* saying that its atoms are that old; indeed, they are much older.

What they are instead doing, when they give the age of a rock, is telling how long ago its constituent atoms became part of that rock.

WE HAVE ARRIVED AT THE STAGE in the development of the early solar system at which there were numerous planetesimals that gravitationally interacted with each other. They might have collided to form one larger planetesimal. Alternatively, a collision could have shattered one or both planetesimals, a setback in the planet-formation process, but only a temporary one, since the resulting debris would, for the most part, have been gobbled up by other planetesimals. It was through this chaotic process that the solar system came to have the planets that it does, and these planets came to have the moons that they do.

When a planet receives a glancing blow from a smaller body, strange things can happen. I mentioned earlier that almost all of the planets rotate counterclockwise on their axis. Uranus is an exception: its axis, rather than pointing "up," points "sideways." This is thought to be the result of a glancing blow by an Earth-sized planet. Mercury is also thought to have been the recipient of a glancing blow that knocked off much of its crust. Since lighter material would have floated up into that crust, Mercury ended up with far fewer light elements than our planet has.

The earth also seems to have been the recipient of a glancing blow. Astronomers think that early in its existence, the earth was struck by a Mars-sized planet. (Mars has a diameter only one-half that of the earth.) Much of the earth's crust was thereby blown into orbit around the earth, where it mixed with material from the impacting body. This material subsequently coalesced to form our moon. This explains why moon rocks are so similar to Earth rocks: many moon rocks used to be Earth rocks. This collision, by the way, would have liquefied most or all of the earth's rocks,[5] thereby resetting the earth's "oldest rock" timer: it is only after magma cools that crystals can reform and their atomic clocks can start ticking.

The creation of the moon has played an important role in our own existence. The impact that gave us our moon probably also gave the earth's axis its 23.5-degree tilt. This tilt, by giving us seasons, makes more of the earth's surface habitable than would otherwise be the case. Furthermore, the presence of the moon stabilizes the tilt of the earth's axis, thereby stabilizing the amount of seasonal variation in temperature. And finally, without the moon, there would still be sun-caused tides, but they would be much lower than those caused by the moon. Tides, as we have seen, probably played an important role in allowing animals to move from the sea to the land. This is because tides create tidal zones in which marine species can "practice" land living by flipping from one tidal pool to another.

About half a billion years after the moon-causing collision, the earth apparently experienced what is known as the *Late Heavy Bombardment*, which might have been triggered by changes in the orbit of Jupiter. During this period, the moon was pummeled by asteroids. The moon's crater-scarred surface is evidence for this: there aren't just craters, but craters within craters within craters. The earth was hit just as hard, but our craters have weathered away. Most likely, life hadn't yet arisen when this bombardment started. It is also possible, though, that it had arisen but was driven to extinction by the bombardment, or that it had arisen and only a few very hardy microorganisms made it through alive.

Since the Late Heavy Bombardment, meteoroids, asteroids, and comets have continued to strike the earth, but in a sporadic manner. Sixty-five million years ago, as we have seen, an asteroid struck the earth on the edge of what is now the Yucatan Peninsula of Mexico and in doing so played an important role in the extinction of the dinosaurs. Fifty-thousand years ago, an asteroid hit what is now Arizona, causing Meteor Crater. In 1908, an asteroid or comet exploded over Siberia, in what is known as the Tunguska event. It flattened trees across a 770 square mile (2,000 square kilometer) area. And in 2012, the Chelyabinsk meteor caused a sensation by passing over Russia.

It is estimated that the earth is hit daily by 100 tons (91,000 kilograms) of cosmic material. Some of it makes it to the surface of the earth, either as dust or as objects that can subsequently be found, most conspicuously, in Antarctic ice and on the roofs of buildings.[6] The rest of the material "burns up" in the atmosphere, maybe as a shooting star. It is important to realize, though, that even though this last material may "disappear," its constituent atoms don't cease to exist. To the contrary, they take up residence in the atmosphere before moving on to become part of the earth's crust, its oceans, or maybe even a living thing—including yourself.

By now, though, it shouldn't be surprising to hear that *some* of your atoms have a cosmic origin; they *all* do. Some might have arrived in a shooting star that a child last year wished upon. Others might have been transported to Earth by the dinosaur-killing asteroid of 66 million years ago. And many more arrived much earlier than that, as passengers on the various cosmic objects that combined to make our planet.

EARLY IN ITS EXISTENCE, AS WE HAVE SEEN, the earth would have been molten. During that period, its elements sorted themselves out, with the heaviest, like nickel and iron, sinking to the earth's core, and the lighter ones rising to form its crust. Consequently, when deposits of heavier metals are found on the surface of the earth, the assumption is that they were brought here relatively recently by an asteroid. This is how geologists think that the nickel deposits found near Sudbury, Ontario, came to be. The gold that we mine from the earth's surface is also thought by some to have come here in late-arriving asteroids.[7]

As the earth cooled, a crust would have formed in much the same way as ice forms on polar seas during the winter. At first, there would have been islands of crust floating in a sea of molten material. These islands would have grown until finally, there would have come the day when the earth was entirely crusted over. (Well, almost entirely: there are still places on the planet with visible lava lakes, including the one in Mount Nyiragongo in the Democratic Republic of the Congo.) Since that day, the crust has been thickening, but very slowly, inasmuch as the crust itself acts as a kind of insulator that hinders further cooling. Even after billions of years, the earth's crust is only 3–5 miles (5–8 kilometers) thick under the oceans and perhaps 25 miles (40 kilometers) thick under the continents. To put this thickness into perspective, we need to keep in mind that our planet is 8,000 miles (12,900 kilometers) in diameter. In proportional terms, the shell of a chicken egg is vastly thicker than the crust of the earth.

The intense heat at the center of the earth causes material to rise. On reaching the crust, it moves laterally and cools. As the result of this cooling, the material becomes increasingly dense until it finally sinks, thereby completing the convection cycle. The lateral current caused by this process transports the tectonic plates that comprise the earth's crust. When two plates collide, something has to give. What usually happens is that one plate is pushed up to create mountains, while the other is pushed down. The plate that has been pushed down will partially melt, and material from it will float up to feed volcanos. This is why the borders between colliding plates are often marked by chains of volcanos, like those on the west coast of South America. Collisions between plates can also cause continents to rise.

The motion of the earth's tectonic plates affects life in many ways. By splitting apart continents or giving rise to mountains, it divides populations of a species and thereby launches them on different evolutionary trajectories. It can also change regional climates, again affecting the evolution of the species that live there. Significantly, plate tectonics is probably responsible for the change in Africa's climate that 7 million years ago transformed rainforest into savanna and thereby set the stage for our evolutionary ancestors to develop the ability to walk upright.

THE MOTION OF THE EARTH'S TECTONIC PLATES has affected life in other ways. Most significantly, were it not for their motion, the earth would likely be a water world, with no land and therefore no land animals. Here's why.

Gravity, as we have seen, is a planetary leveler: it is what makes planets round. When water is present, though, the leveling process moves into high gear. In its liquid form, water washes things downhill, and in the form of ice-bearing glaciers, it scours the countryside and pushes the resulting debris downhill. Furthermore, by changing from liquid to ice and back again, water has the ability to break cliffs into boulders and break boulders into rocks that can be transported and further broken down by rivers. Working together, gravity and water have the power, given enough time, to tear down mountains. For an example of this phenomenon, consider again the mountains, described in chapter 6, that once stood where the island of Manhattan now is. Although those mountains were thousands of feet high, Manhattan's current highest (natural) elevation is a mere 265 feet (80 meters) above sea level.

The process of erosion moves material from land to the sea, where it is deposited, at the end of rivers, as deltas and further offshore as fans. The Bay of Bengal, between India and Myanmar, has been receiving material eroded from the Himalayas for maybe 70 million years. The resulting submarine formation, known as the *Bengal Fan*, is nearly 2,000 miles (3,200 kilometers) long and 600 miles (1,000 kilometers) wide, and in some places is 10 miles (16 kilometers) thick. The continental shelves that skirt continents are also made up of material eroded from the land.

In much the same way as the level of water in a bathtub rises when you sit down in it, the level of the oceans rises when eroded material is deposited in them. This means that at the same time as erosion is reducing the height of the land, it is raising the level of the sea, and as a result, coastal areas are submerged. This in turn means that if the earth's mountains were not constantly being rebuilt by the forces of plate tectonics, all land would ultimately be submerged, and Earth would become a water world.

And this isn't just something that *could* happen. As the earth continues to cool, its crust will thicken so much that its tectonic plates can no longer move. At that point, mountain building will be much reduced,[8] but the forces of erosion will continue. The combination of shrinking mountains and rising oceans will ultimately transform the earth into a world without dry land. The ocean that covers it will be, on average, 9,000 feet (2.7 kilometers) deep.[9] Water worlds of this kind, I should add, aren't merely theoretical entities. Jupiter's moon Europa and Saturn's moons Titan and Enceladus are believed to have deep oceans under their surface ice.

This brings us to the question of where the earth got its water. It obviously came here the way all the earth's atoms did, from our surrounding cosmic environment. The obvious sources would be impacts by icy comets and watery planetesimals. Realize, though, that even a seemingly dry asteroid could carry "chemically bound" water that would be released when the asteroid was heated by its collision with the earth. And in thinking about the earth's water, it is important to keep in mind that in the same way as nickel and iron will naturally settle to the earth's core, water will naturally rise to its surface. If it is carried deep underground, it will be heated, turn to steam, and be released at the surface. As a result, much of the earth's water is at its surface.

Because the earth's water is readily visible in lakes and oceans, it is easy to form the impression that our planet has lots of the stuff, but this impression is very much mistaken. By volume, the earth is only 0.128 percent water, and by weight, it is an even lower percentage than that, since minerals are almost always heavier than water. Gather up all this water into one drop, and it would be only 860 miles (1,380 kilometers) in diameter, not even enough to span the United States (see Figure 16.3).

SO MUCH FOR THE EARTH'S OCEANS. What about its atmosphere? Our exploration of the solar system has shown us that many different planetary atmospheres are possible. Mercury is baked by the sun and blasted by the solar wind and, as a result, has almost no

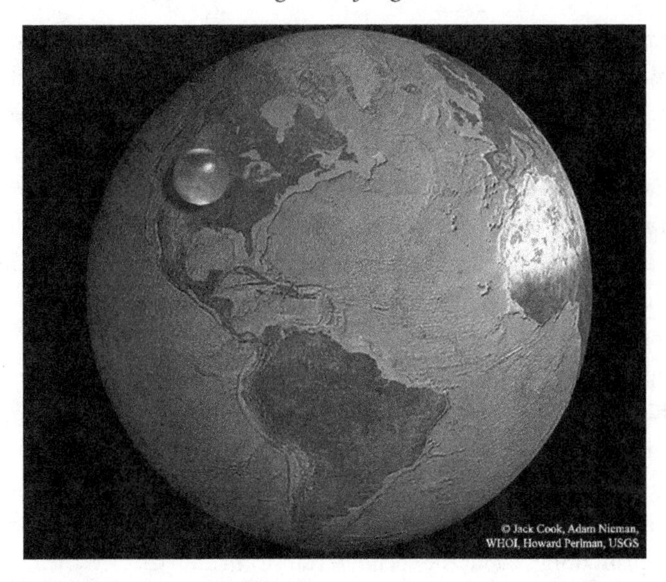

FIGURE 16.3. This is what the earth's water would look like if it were gathered into one big drop, leaving the rest of the planet dry.

atmosphere. Venus's proximity to the sun would make it hot, but its atmosphere makes it hotter still: its atmosphere is not only thick but is 95 percent carbon dioxide, making it subject to an extreme form of the greenhouse effect. As a result, daytime temperatures of 800 degrees Fahrenheit (425 degrees Celsius) are common. Mars has a much thinner atmosphere than the earth—it was blown away, as I've said, by the solar wind—and what little atmosphere remains is 95 percent carbon dioxide. The gas giants Jupiter and Saturn, by way of contrast, have very dense atmospheres with a chemical composition similar to that of the sun, which is to say that they are about 75 percent hydrogen and 25 percent helium. They also have small amounts of methane, ammonia, hydrogen sulfide, and water.

The earth's atmosphere is thicker than that of Mars or Mercury, but much thinner than that of Venus, and very much thinner than that of the gas giants. It is at present 78 percent nitrogen, 21 percent oxygen, and 1 percent argon, with trace amounts of water vapor and carbon dioxide, but this has not always been the case. Before there were cyanobacteria, for example, the earth's atmosphere contained very little oxygen, but by 300 million years ago, oxygen levels had risen to over 30 percent. And as a result of the industrial revolution, the amount of carbon dioxide in the atmosphere has risen from 300 parts per million (0.03 percent) to 390 parts per million.

The composition of a planet's atmosphere is determined by the rate at which gases are added to it and the rate at which they are removed. Consider argon, for example. Many people are surprised to hear that it is the third most common gas in our atmosphere, behind nitrogen and oxygen, but its presence there is easy to explain. It is a product of the radioactive decay of potassium-40, which is common in the earth's crust. It is also a noble gas, meaning that once it gets into the atmosphere, it doesn't combine with other elements to form compounds that could be removed from the air. Finally, it is heavy, meaning that it doesn't drift into the upper atmosphere, where the solar wind could carry it away. Consequently, once it gets into the atmosphere, it tends to stay there.

You might assume that nitrogen's abundance in the atmosphere is a consequence of it being abundant on the earth, but you would be mistaken. The earth has precious little nitrogen: in the list, in order of abundance, of the elements in the earth's crust, nitrogen comes in #32, just behind yttrium and scandium, which are hardly household names. Although nitrogen makes up 78 percent of the earth's atmosphere, it makes up a scant 0.002 percent of the earth's crust.[10]

It would appear, then, that much of the earth's nitrogen is in its atmosphere. It got there, geologists think, as the result of plate tectonics. When rock containing nitrogen-bearing compounds is subducted under the surface of the earth, heat liberates N_2 molecules, which are then vented by volcanos into the atmosphere.[11] Because N_2 molecules are, as we have seen, relatively inert, they tend to stay in the atmosphere unless lightning, bacteria, or a fertilizer plant brings about the chemical transformation that lets them return to the earth's crust.

The atmosphere's oxygen, in the form of O_2 molecules, has a rather different source—namely, plants on land and, more significantly, phytoplankton in oceans. Realize, though, that unlike N_2 molecules and argon atoms, O_2 molecules are chemically promiscuous entities: left to their own devices, they will combine with other atoms to make molecules, including H_2O and CO_2, that in the natural course of things will be removed from the atmosphere. Bottom line: take away the earth's plants and phytoplankton, and its atmospheric oxygen goes with them.

We have seen the role plate tectonics plays in keeping portions of the earth above sea level and thereby making land possible. Without atmospheric oxygen, though, that land might have been uninhabitable. When O_2 molecules drift high into the atmosphere, they are struck by UV rays that transform them into O_3 molecules, also known as *ozone*. These molecules subsequently filter out the UV rays that would otherwise make life very difficult for land animals.

Earth not only has atmospheric oxygen but has what is, for us humans, an excellent amount of the stuff—21 percent. If we had much less, we would feel enervated, the way we do when we travel to high altitudes. It is, to be sure, possible to adapt to reduced oxygen the way Tibetans have—thanks, as we saw in chapter 6, to their Denisovan ancestors. Reduce the atmosphere's oxygen content below 16 percent, though, and fires will no longer burn. Without fires, our ancestors wouldn't have been able to cook. Deprived of cooked food, the human brain probably wouldn't have grown the way it did, meaning that we humans would not have developed, intellectually and culturally, the way we have.

Suppose that instead of being lower, the oxygen content of the atmosphere had been rather higher. Under those circumstances, fires would be easy to start but on starting, would voraciously consume whatever fuel was available. Lightning strikes, rather than causing small fires that our ancestors could exploit, might instead result in forest-consuming conflagrations. In such an atmosphere, we humans, rather than harnessing fire, might quite reasonably have lived in fear of it.

HERE, IN A NUTSHELL, IS YOUR ATOMIC HISTORY. Some of your atoms were created shortly after the Big Bang. Others were created in stars that subsequently exploded. These last atoms ended up in a cloud of supernova debris. Part of that cloud subsequently collapsed to form the solar system of which the earth is a member.

Some of these earthly atoms ended up in H_2O, CO_2, O_2, and N_2 molecules that, as we have seen, blew in the wind. They subsequently might have spent time in some living thing—in a plant, for example, or a plant-eating animal. It is also entirely possible that some of your atoms had previous human owners, or that they were previously inside the gas tanks of cars or the bladders of ancient dinosaurs. You might think that becoming part of you is the most interesting thing that ever happened to your atoms, but your atoms, if they were capable of having a sense of humor, would find the idea laughable.

PART IV

Your Place in the Universe

17

You Are a Gene Machine

MY WIFE AND I HAVE TYPE B BLOOD. When our son was born, the doctor informed us that he had type O. The room went silent. On detecting our puzzlement, he explained how such a thing could happen. Everyone inherits two blood-type genes, one from their mother and one from their father. Each gene can be one of three variants, also known as *alleles*: A, B, and O. This means that each of us[1] has one of six blood genotypes: AA, AB, AO, BB, BO, and OO. This genotype in turn determines your blood type. In particular, to have type B blood, you must have either genotype BB or BO, and to have type O blood, you must have genotype OO.

It is therefore possible for a child to inherit type O blood from two type B parents, as long as both parents are genotype BO: he gets one O gene from the father and the other O gene from the mother. This would happen, on average, to only one in four children whose parents had the BO genotype. In the other three cases, the child would get his father's B gene and mother's O gene, or his father's O gene and mother's B gene, giving him the BO genotype and type B blood; or he would get his father's B gene and his mother's B gene, giving him the BB genotype and again type B blood. Mystery solved. Whew!

As we have seen, genes are recipes for the construction of proteins. These recipes are written with nucleotides on the strands of your DNA. Biologists refer to these strands as *chromosomes*. You have 46 of them, with 23 from your mother and 23 from your father. Your 20,000 or so genes are not placed randomly on your DNA but in a specific order, on specific chromosomes. For example, the gene that determines your blood type—geneticists call it your *ABO gene*—is found on chromosome #9 at position 34.2 on its longer arm (see Figure 17.1). This chromosome contains 141 million of the 3.2 billion base

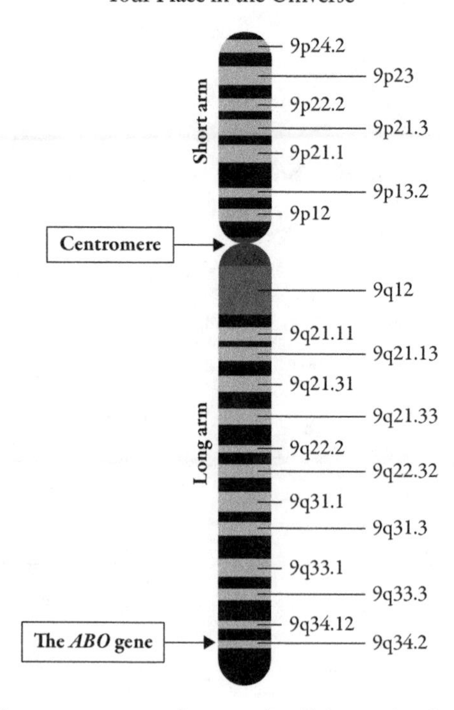

FIGURE 17.1. Human chromosome #9. It is home to the *ABO* gene that determines your blood type.

pairs that comprise your DNA. It is also home to more than a thousand other genes, including *COL5A1*, the gene for one of the several kinds of collagen that your body makes.

Your body routinely makes copies of your DNA. In particular, before a cell fissions, its DNA is copied, so that each daughter cell can have its own DNA. The copying process does not change the order of the nucleotides in those strands, any more than photocopying a document changes the order of the words on the page or the order of the letters in those words. Consequently, if a DNA copy is accurate, it will have the same nucleotides in the same order as the original strand did, and will therefore have the same genes in the same order.

It is tempting to think of your DNA as a cookbook in which gene recipes are stored, but only 1–2 percent of your genome plays this role.[2] The DNA that constitutes the other 98–99 percent plays a variety of roles. There are, for example, stretches of DNA that play a regulatory role, allowing genes to be turned on and off. Other stretches serve as landmarks on the DNA strand. There is also lots of what is apparently non-functional DNA, sometimes dismissively referred to as *junk DNA*. It is likely, though, that much of this maligned DNA does in fact function but in ways we don't yet understand.

Before we move on, some comments are in order. Genes, as we have seen, are recipes for proteins. It turns out, though, that one gene can give rise to multiple proteins, depending on which parts of the gene are expressed and which aren't.[3] Your genes therefore resemble culinary recipes that have options—that might, for example, give a chef the option of adding butter at one point in the preparation process or omitting it. Furthermore, not all of your 20,000 genes are essential for your well-being. Many of them can be removed from your genome without apparent effect.[4]

YOU INHERITED ONE #9 CHROMOSOME from each of your two parents. Instead of simply passing down one of her #9 chromosomes to you, though, your mother's reproductive system "shuffled" her two #9 chromosomes together and then randomly chose one of the two shuffled chromosomes to give you. As a result of this shuffling process, more formally known as *genetic recombination*, the chromosome #9 you inherited from her is different from either of her #9 chromosomes.

The shuffling process is rather complex, but here is one way to understand it. Suppose you had two decks consisting of 100 cards, with the cards in one deck being blue and the cards in the other deck being pink. Suppose that the cards in each deck were numbered from 1 to 100 and were piled in numerical order, with 1 being the top card. Suppose, finally, that you shuffled these decks not in a random manner, but by carefully exchanging randomly chosen sections of cards between them.

You might, for example, randomly choose two numbers between 1 and 100. If you chose, say, 17 and 31, you would switch cards 17–31 in the pink deck with cards 17–31 in the blue deck, being careful to keep the cards in the switched sections in their original order. You would end up with two 100-card decks, and the cards would still be in numerical order, with card 1 on the top and card 100 on the bottom. Each of these shuffled decks, however, would now be a mix of blue and pink cards—in particular, they would have different colored cards 17–31 than they formerly did—meaning that each of the two shuffled decks would be different than either of the two original decks. Notice, by the way, that although the shuffling process affects the two decks, it does not affect individual cards. In particular, pink card number 23 will not itself be changed by being moved to the blue deck.

The chromosome you inherited from your mother consisted only of genes from her two #9 chromosomes, and those genes would be in the same order as the genes in those chromosomes. As a result of genetic recombination, though, the inherited chromosome would have sections of genes that came from her mother's #9 chromosome and sections that came from her father's. They would resemble, in other words, the shuffled pink and blue decks of cards. And in the same way as shuffling those decks does not alter the identity of the individual cards within them, the genetic recombination of her chromosomes does not affect the individual genes within those chromosomes.

Consider, in particular, the gene at position 34.2 on the longer arm of the #9 chromosome you inherited from your mother. At that spot will be either the *ABO* gene that your mother inherited from her mother or the *ABO* gene that she inherited from her father.

The chromosome #9 you inherited from your father is likewise the result of a shuffling process.

Because your parents' reproductive systems shuffled their genes and then randomly selected which shuffled version to transmit, you are genetically different from either of them. And since the shuffling is likely to produce a different result with each offspring, you are genetically different from your full (non-identical-twin) siblings. Yes, they also got half of their genes from your mother, the way you did, but because of shuffling and random selection, they got a *different* half than you did. The same can be said of the genes they got from your father.

There is, however, one important exception to this shuffling process. Your father's reproductive system does not transmit a shuffled version of his chromosome #23, also known as his *sex chromosome*.[5] He instead transmits either the exact X chromosome he inherited from his mother, in which case you are a female, or the exact Y chromosome he inherited from his father, in which case you are a male.

THE PROCESS BY WHICH YOUR PARENTS shuffled their chromosomes before passing them on to you is known as *meiosis*, and it is simultaneously quite complicated and incredibly beautiful. I have long regarded it as a double miracle, in a non-theological sense of the word. The first miracle is that nature could have "invented" such a process, and the second miracle is that biologists could subsequently have figured it out. Furthermore, the phase of meiosis known as *crossing over* is of particular interest, inasmuch as it is, if described with a bit of imagination, the most erotic process in all of biochemistry. The story I am about to tell takes some artistic license, but I suspect that you will find it more memorable than the description of crossing over that you heard in your high school biology class.

In order to make an egg, a cell in your mother's ovaries made a copy of the #9 chromosome she inherited from her mother. After being copied, these identical chromosomes—now referred to, somewhat confusingly, as *sister chromatids*—were joined at their centromeres and therefore resemble the letter *X*. The center of the *X*, where the two centromeres were connected, could be thought of as the belly button of the *X*, with the branches of the *X* resembling arms and legs. The ovarian cell also did this with the #9 chromosome she inherited from her father. As a result, there were two #9 *X*s, one maternal and one paternal.

These two *X*s found each other and came together at their "belly buttons." They entwined the "arms" and "legs" of their respective *X*s and—what else?—exchanged genetic material. They might have swapped a "knee" or a "hand," for example, thereby producing new, shuffled chromosomes. The pair of shuffled *X*s, exhausted and yet deeply satisfied, separated. Each shuffled *X* then separated into its two component chromosomes. This intracellular sex happened, by the way, not just before your mother laid eyes on your father but before she was born.

In the end, there were four chromosomes, no two of which were alike, and none of which were the same as your mother's own maternal and paternal #9 chromosomes. One of these four chromosomes went into the egg that became you. And although I have focused my attention on the #9 chromosome for the telling of this story, it was by no means special. Your mother's other chromosomes underwent the same process.

To make a sperm, the chromosomes in one of the cells in your father's testes similarly engaged in intracellular sex. This would have taken place about three months before the interpersonal sex that resulted in your conception. In the maybe four decades between menarche and menopause, your mother might have had a total of 500 eggs in play. Meanwhile, your father, starting at puberty, would have been producing more than 100 million sperm a day. That works out to more than 1,000 per second and more than 2 trillion in the course of a lifetime![6] This means that after puberty, a male is faced with a serious and ongoing sperm disposal challenge—one, I should add, that he meets bravely and without complaint.

IN MUCH THE SAME WAY AS WE CAN CONSTRUCT a genealogical tree to trace your ancestry, we can, for each of your genes, construct a genetic tree that indicates which of your ancestors it came from. This tree will therefore track along your family tree: it will show, for any given gene, which of your parents it came from, which of your parents' parents it came from, and so on. As we have seen, you have two copies of the *ABO* gene. Figure 17.2 shows how the tree for these copies might look. One gene came to you from your father and before that, from your father's ancestors. The other gene came from your mother and before that, from your mother's ancestors.

If you are a male, the tree for a gene on your Y chromosome will look rather different. Because Y chromosomes are transmitted, without alteration,[7] from father to son, you can trace the genes on your Y chromosome back through your patrilineal ancestry—through

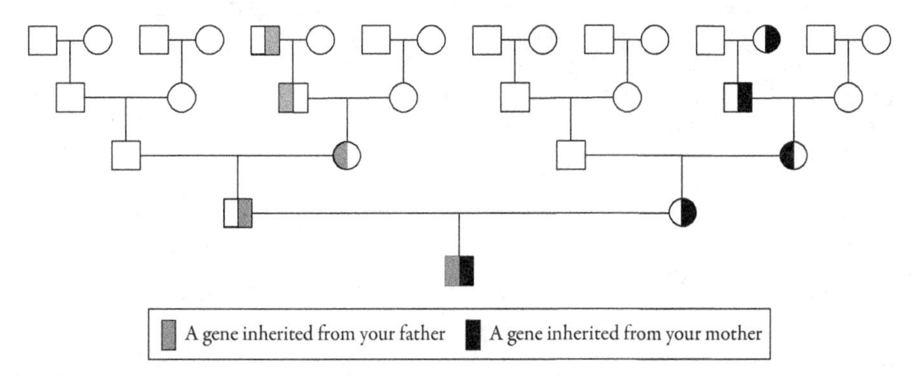

FIGURE 17.2. A "genetic family tree" that traces your two *ABO* genes back through your ancestry. The one shown in gray is the allele you inherited from your father, and the one shown in black is the allele you inherited from your mother.

your father, your father's father, and so on (see Figure 17.3). The genes on your X chromosome, however, cannot likewise be traced back through your matrilineal ancestry. This is because her reproductive system created the X chromosome that you inherited by shuffling together the two X chromosomes that she had inherited from her parents.

To be thorough in our genetic history of you, we will have to provide trees not only for the genes in your nuclear DNA but for the genes in your mitochondrial DNA. Since you inherited your mitochondria from your mother, you inherited your mitochondrial genes from her as well. This means that the tree for one of these genes can be traced back through your matrilineal ancestry—to your mother, your mother's mother, and so on. Go back far enough, and you will arrive at *Becky*, the Alphaproteobacterium, described back in chapter 11, that is the ancestor of all mitochondria.

WHEN WE CONSTRUCTED THE TREE OF LIFE, we saw that any two currently existing species have an ancestor in common; the only question is how far back we have to go to find that ancestor. Since two currently existing species will have inherited the genes of their common ancestor, they will have genes in common.

We have seen that genes are not themselves altered by the genetic recombination that takes place during genetic reproduction. From this, however, it does not follow that genes are unalterable. Exposure to mutagens, including radiation and various chemical substances, can transform them. They can also be changed by mistakes made when DNA strands are copied. Given enough time, sufficiently many genes can be sufficiently altered that the original species will have transformed into a new species. There is, however, a random component to these genetic changes. Consequently, if members of one species are split into two groups that don't interbreed for an extended period, there is a good chance that genetic drift will transform the two groups into two different species.

This is presumably what happened to the common ancestor of humans and chimpanzees. And because this ancestor lived "only" 7 million years ago, the genetic difference between us is surprisingly small: it has been estimated that we and chimpanzees

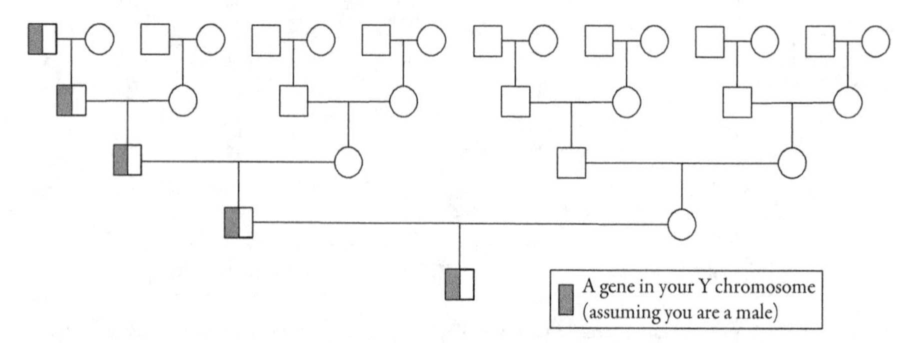

FIGURE 17.3. If you are a male, your Y chromosome and the genes it contains can be traced back through your patrilineal ancestors—your father, your father's father, and so on.

have about 99 percent of our DNA in common.[8] Generally speaking, the further back we have to go to find the ancestor we share with another species, the less DNA we have in common.

Despite our genetic similarity to chimpanzees, we are markedly different from them in one important genetic respect: whereas we have 23 pairs of chromosomes, they have 24. But how can two organisms that share a common ancestor only 7 million years ago come to have a different number of chromosomes? It would appear that either humans lost a chromosome or chimpanzees gained one, but which was it? And as long as we are asking questions, wouldn't a 4 percent difference in the number of chromosomes mean a 4 percent difference in our genes, rather than the 1 percent mentioned earlier?

It turns out that we humans "lost" a chromosome, but in a surprising way.[9] The ancestor we share with chimpanzees presumably had 24 chromosomes, the way modern chimpanzees and other great apes do, but during the next 7 million years, two of those chromosomes fused to make what we call our chromosome #2. (If you carefully examine the structure of this chromosome, you can find the point of fusion.) In place of our chromosome #2, chimpanzees have two chromosomes, #2A and #2B. Between them, these two chromosomes have nearly the same DNA as our chromosome #2 does. This is why, although we "lost" a chromosome, we didn't lose the genes that were written on it.

I HAVE ALREADY GIVEN A RUDIMENTARY DESCRIPTION of LUCA, the most recent ancestor that all living things have in common. It was a single-celled organism that used the "universal" genetic code to construct proteins in accordance with the instructions provided by codons of DNA. Whereas modern humans have 3.2 billion base pairs in their DNA, LUCA probably had fewer than a million, and whereas modern humans have 20,000 genes, LUCA probably had a few hundred,[10] and the earliest living things doubtless had fewer still. This raises the question of where our additional genes came from.

We have seen that copying errors can alter a particular gene. They can also result in a gene getting copied twice, through the process known as *gene duplication*. One of those copies might subsequently mutate or be miscopied, meaning that it will be different from its twin gene. One gene will thereby become two.[11]

Another way an organism can get new genes is from an external source. Bacteria accomplish this by acquiring from their neighbors the little rings of DNA known as *plasmids*. Such transfers can be beneficial to bacteria; indeed, it is one way antibiotic resistance spreads among them. Because of this *horizontal gene transfer*, as it is called, the genetic tree for a bacterial gene will have "horizontal branches."

Humans can also acquire genes from an outside source, via viruses. In most of the cases in which you are infected by a virus, it takes control of your body's gene-copying apparatus to make copies of itself, while leaving your own DNA untouched. Sometimes, though, a virus splices genes into your DNA. If the attacked cells are somatic cells, these spliced genes perish when you die, but suppose they are instead germ-line cells. If these cells subsequently produce egg or sperm cells, and if these gametes are part of a successful reproductive act, the resulting organism will carry the viral DNA as part of its genome. This phenomenon, we need to realize, isn't just possible it is relatively common. We have seen that less than 2 percent of our DNA codes for genes. Another 8 percent is the result of ancient viral attacks.[12]

It is easy to experience a feeling of resentment on hearing of these attacks: how dare viruses alter our genome! But realize that these attacks aren't necessarily a bad thing. One such attack is thought to have provided us with the gene responsible for the production of the protein syncytin.[13] This protein—actually, it is a group of related proteins—allows a placenta to fuse to the lining of the uterus. Without it, you wouldn't have been able to develop in the sheltered environment of your mother's womb. And this isn't the only way syncytin benefits us. We have seen that in the normal course of things, cells divide, with one cell becoming two. Syncytin, however, allows two muscle cells to fuse into one,[14] thereby increasing their strength.

It should be clear from this discussion that extended genetic trees will be dramatically more complex than the one shown in Figure 17.2. Genes can be altered by mutation and copying errors. Members of one species can acquire genes from members of another. (Your single-celled ancestors probably did lots of this gene swapping.) And in surprisingly many cases, we will be able to trace a gene back to a virus. Furthermore, it is likely that genetic trees are going to become even more complex in the future: it isn't long, one imagines, before humans will be able to trace some of their genome back to "artificial genes" that were created by genetic engineers and implanted into the human germ line. As a result, the genetic tree of one of your descendants might have a decidedly improvised look.

It should also be clear that it is misleading to speak of *the* genome of a species. It is misleading, in particular, to speak of *the* human genome, inasmuch as no two humans (other than identical twins) have the same genome. Furthermore, the genomes of members of a species will change with the passage of time, as genes come, genes go, and genes transform into different genes.

We living organisms are in a battle for survival. Those of us who are well suited to our environment survive and reproduce; those who are not, perish. Our genes are likewise in a battle for survival, and their struggle parallels our own. If a change in a gene makes it easier for members of a species to survive and reproduce, the new gene will spread, and if the change instead makes it harder for members to survive and reproduce, the new gene will disappear. Besides thinking of the evolution of species, then, we should think in terms of the evolution of genomes. In fact, we might draw the conclusion that the evolution of genomes is more fundamental than the evolution of species—that the later phenomenon is little more than a manifestation of the former.

YOU DOUBTLESS THINK OF YOUR GENES as being, well, *your* genes, but do they really belong to you, the way that, say, your atoms do? To better understand the sense in which your atoms are yours, suppose someone started removing them without your permission. They would be guilty at least of battery, since the removal would involve unwanted touching, and they would be guilty of an even worse crime if it involved the removal of, say, your left kidney. Conclusion: you legally own your atoms. Other people have atoms that are chemically identical to yours, but the atoms in question are physically distinct from yours and clearly belong to them instead of you.

When it comes to your genes, though, the question of ownership gets complicated. As we have seen, you inherited one of your *ABO* genes from your mother, and barring mutation, your *ABO* gene is "genetically identical" to one—or maybe both—of her *ABO* genes. The gene in question will be written into your DNA: it will be represented there by a sequence of nucleotides. Although you clearly own these nucleotides, a case can be made that you don't likewise own "your" *ABO* gene. What you instead possess is a *copy of* the *ABO* gene. So does your mother, along with many other people.

To better understand this point about ownership, consider Robert Frost's poem "The Road Not Taken," the first line of which is "Two roads diverged in a yellow wood." I have a copy of this poem. So do many other people, and unless copying mistakes have been made, their copies are exactly the same as mine, which in turn is exactly the same as Frost's original copy. Yes, the fonts may be different, and they are printed on different pieces of paper with different ink, but the key thing is that they contain precisely the same information. Therefore, they are, for literary purposes, identical copies of the poem.

Suppose a literary vandal managed to destroy all the physical copies of "The Road Not Taken." Electronic copies of the poem would still exist. Suppose this vandal subsequently erased all such copies. As long as there were people who had memorized the poem, it would continue to exist, in mental form. And a case can be made that even if these individuals all died, taking their memories with them, the poem would not thereby cease to exist. It would instead be *lost*, the way Homer's epic poem *Margites* is. Such a loss, however, would leave open the possibility that some future poet would independently create Frost's poem and put it down on paper as her own creation. And along these lines, realize that Frost himself might not have created "The Road Not Taken" but might simply have unwittingly "rediscovered" a poem written by some other poet.

Much the same can be said of your genes. You may own the medium on which your *ABO* gene is written—namely, a stretch of the DNA in your #9 chromosome—but the gene itself can best be thought of as information that exists independently of any particular record of it. If you die and your DNA subsequently degrades, your copies of the *ABO* gene will cease to exist, but other people's copies will continue to exist. Furthermore, if these other people all die, the *ABO* gene can continue to exist, in the form of a sequence of nucleotides recorded in a computer file. And finally, if all these files are deleted, it would be incorrect to say that the *ABO* gene had ceased to exist. It would more properly be described as a gene that had been lost to living things, leaving open the possibility that it might someday be "rediscovered" by evolutionary processes—or reinvented by a clever geneticist.

IT IS SOMEWHAT DISCONCERTING TO REALIZE that you don't own your genes. A case can be made, though, that the situation is even worse than this: you are being used by those very genes! Biologist Richard Dawkins is the most famous advocate of this selfish-gene theory.[15] Genes are not, to be sure, consciously selfish—they can't be since they don't have minds—but they certainly act as if they were. From their (imaginary) point of view, your job is to make copies of them, and they accomplish this by taking control of your mind.

To begin with, they have "wired" your brain so that you have a survival instinct. You are wired, for example, to struggle for air when deprived of it, to experience a fight-or-flight response when confronted by a bear, and to seek food when you haven't eaten in a while. You are also wired to have a sex drive that makes you seek a mate. This wiring increases the chance that you will survive and reproduce—and thereby make copies of your genes. And finally, you are wired so that you are willing to sacrifice on behalf of the children you bring into existence; indeed, you are so thoroughly wired that the sacrifices won't feel like sacrifices. This wiring further increases the chance that "your" genes, carried by your offspring, will outlast you.

For an extreme example of the way genes can affect behavior, consider social insects. Worker bees spend their lives slaving away for the good of the colony without any chance of themselves reproducing.[16] At first this sounds like a foolish thing to do, from a genetic point of view. But when we recall that worker bees share the genes of the queen, we realize that by helping her reproduce, they are propagating "their" genes.

The situation of your somatic cells resembles that of worker bees. These cells spend their lives slaving away for the good of your body even though their cell lines will terminate when you die. This sounds sacrificial until we recall that the genes in your germ cells are also present in your somatic cells. Consequently, by working to increase the chance that your germ cells will make the gametes that allow you to reproduce, your somatic cells are increasing the chance that "their" genes will propagate.

If your genes could talk, they might explain to you just how insignificant you are, seen from their point of view. Whereas a person like yourself might, if lucky, live for a hundred years, and a tree might, if lucky, live for thousands of years, lots of genes have been around for millions or even billions of years. They might add that although your participation in their propagation efforts is appreciated, it is by no means essential. In particular, even if you had never existed, "your" genes would almost certainly have been propagated by other people and in many cases by other living organisms.

You may resent being used in this manner by your genes, but you will have to admit that as exploitation strategies go, it has been remarkably successful.

18

You Are (Merely) Part of Life

YOU ARE, OF COURSE, ALIVE, or you wouldn't be reading these words. The cells that comprise you are also alive, as are the non-human cells that live in and on you. There are also living things in your immediate environment, including whatever people you share your home with, along with whatever pets and plants you own. But this is only the beginning of your domestic census.

For one thing, you also share your home with very many "bugs," including spiders, flies, beetles, and centipedes. One survey of 50 homes in and around Raleigh, North Carolina, discovered that the average house sheltered 93 distinct kinds of arthropod. The least inhabited home had 32 different kinds, and the most inhabited had 211.[1]

My wife and I share our home with what, according to the aforementioned survey, would appear to be a typical assortment of bugs. A decade ago, I regarded these bugs as intruders, as do most people, but then a minor outbreak of ants in my kitchen changed my mind. My wife instructed me to "do something" about them, and so I went to the library in search of information on how to get rid of ants. In the course of my research, I came across Bert Hölldobler and Edward O. Wilson's magnificent book *The Ants*. As a result of reading it, I gained a profound appreciation for ants—truly, they are marvelous little creatures. This in turn triggered in me an interest in and appreciation of household bugs in general.

Consider centipedes. They are the biggest of the indoor bugs found in my part of the world. Including their legs and antennae, they might be two inches (five centimeters) long, and thanks to the remarkable coordination of their many legs,[2] they are capable of moving with surprising speed. The combination of size and speed makes them a bit scary, but a person who dislikes bugs should not give in to this fear and kill resident centipedes.

They are, after all, high up on the bug food chain and spend their days eating other, smaller bugs. How else could they get that big? By similar logic, if you dislike bugs, you should like spiders, inasmuch as they make their living killing and eating other bugs.

On gaining an appreciation of the bug world, I ceased to regard the occasional visiting arthropod as a problem. Instead, I found myself taking interest and even delight in these little creatures. Some of my enthusiasm has rubbed off on my wife. As a result, her instinct on seeing a bug is no longer to kill it on the spot but to take it outside, to what she believes is its proper domain. On finding a centipede, for example, she will put a plastic cup over it, slide an index card under the cup, and transport the lucky bug to the front yard. Brave woman!

Of course, for every bug your house contains, it has billions of microbes. They are in the air you breathe and on every visible surface. This means that even if you move out of your home, take your pets and plants with you, and exterminate all its bugs, your house will still be full of life. But not to worry, since as we have seen, only a fraction of 1 percent of those microbes are capable of making you sick, and your body has multiple lines of defense to prevent them from doing so.

WALK OUT YOUR FRONT DOOR, and you will discover even more life. Start by examining your lawn, if you have one. Extract a one-cubic-foot sample of soil from it, and you will find a dazzling array of living things. On the top will be grass and maybe some weeds, and running through the cube might be the roots of nearby bushes and trees. You will also find bugs—including, perhaps, some that formerly lived in your home—as well as earthworms, nematodes, fungi, and a vast array of microbes. It is quite likely that by very carefully examining this cube of soil, you will discover a new species of microbe—in your own front yard, no less! And if you instead very carefully examine the soil of a nearby meadow, you are almost certain to discover many new species.[3]

The animate nature of soil becomes apparent when it rains. This is when we are treated to that delicious wet-soil smell, the source of which is primarily geosmin, an organic compound produced by soil-dwelling actinobacteria.[4] We humans are acutely sensitive to its scent: dissolve one teaspoon of geosmin in 200 Olympic-sized swimming pools, and we will still be able to smell it. It would appear, though, that camels are even more sensitive. It is thought that they rely on the scent of wind-wafted geosmin to track down water that is 50 miles distant.[5]

That life would be found in topsoil isn't surprising, but it can also be found in the subsoil and, more impressively, in the bedrock under the subsoil. Bacteria have even been found thousands of feet down in the tunnels of gold mines—more precisely, in the water that has invaded the cracks in the rocks through which the tunnels were dug. Some of those bacteria have figured out how to make a living by chemically combining the hydrogen emitted by the radioactive decay of the rocks with the sulfur contained in those rocks.[6] And it isn't just bacteria that have been found deep underground; nematodes, which are animals, have been found in the water-filled fissures of rocks 2.2 miles (3.6 kilometers) deep.[7]

The earth's oceans are even more alive than its land. Whales have been found 2.3 kilometers (1.4 miles) down, crustaceans 7 kilometers (4.3 miles) down, and fish 8 kilometers (5.1 miles) down. Eukaryotic foraminifera have been found 11 kilometers (6.8 miles) down, in the deepest part of the ocean. Bacteria can be found at all depths. Furthermore, besides being found in the ocean, life can be found under it, in subseafloor sediments. The organisms discovered there have metabolisms so slow that they can easily be taken for dead.[8]

The earth's atmosphere is also full of life. Birds have been observed flying as high as Mount Everest. And besides the obvious birds, bats, and insects, the atmosphere carries an abundance of microbes. They have been found in the upper troposphere, where they play an important role in the earth's weather: they serve as nuclei around which water vapor molecules can condense and thereby trigger the development of water droplets and ultimately clouds.[9]

Besides being widespread, life on Earth is hardy. It can, in particular, tolerate a wide range of temperatures. Life has been found not only under snow, in the ecosystem known as the *subnivium*, but under thousands of feet of Antarctic ice.[10] Other living organisms can withstand extreme heat. Before 1966, researchers assumed that there was no point in looking for life in hot springs, since no living thing could tolerate their heat. In that year, though, microbiologists Thomas Brock and Hudson Freeze took a closer look at the water in the hot springs of Yellowstone National Park and, much to their and everyone's amazement, found living bacteria. Life can also be found around the underwater volcanic vents that pump out water so hot that it would boil, were it not under immense pressure as the result of its depth. But then again, perhaps we shouldn't be surprised to find living organisms in such a place, inasmuch as many scientists think that life on Earth arose in and around similar vents.

Microbes can also live in what we would regard as chemically inhospitable environments, including places that are extremely salty, extremely acidic or basic, or highly radioactive. Consequently, what to us counts as a cave full of toxic sulfurous gas might, to a microbe, count as home, sweet home.[11] Life needs water, so you would think that it could not be found in the extremely dry Atacama Desert in Chile. And indeed, at the surface, you won't find any plants or animals. Dig down, though, and you will find microbes.[12] Along similar lines, although we humans would regard crude oil as a toxic brew, there are microbes for which it is a life-giving broth. And while a decaying lithium-ion battery might not seem like a propitious place to make a living, for some organisms it is the microbial equivalent of an all-you-can-eat buffet.[13]

There may be places on the surface of the earth where life cannot be found. Because it is simultaneously very cold and very dry, for example, University Valley in Antarctica is apparently devoid of life[14]—assuming, of course, that we don't include the scientists doing research there, along with their microbiomes, in our census of living things. Likewise, the inside of an autoclave that has just completed its sterilization cycle will be devoid of life—until the door of the autoclave is opened, at which time airborne microbial life will rush in.

Because Earth-based life is so opportunistic, we have to take care, in our investigations of other worlds, not to inadvertently contaminate them. Indeed, as we saw in chapter 8, we ourselves might be the result of an accidental contamination of the earth by alien beings.

IT IS POSSIBLE FOR THE EARTH to be inhabited by a single species; otherwise, life couldn't have arisen. Subsequently, though, the existences of the various species have become intricately connected. This means that we cannot give a comprehensive description of life on Earth simply by listing the species that inhabit our planet. We must also describe the ecosystems in which those species play a role. It would appear that beyond a certain point of development, life needs life.

This is most dramatically true at the microbial level. Zoologists can easily pick out one species of animal and isolate it from others for further study. When microbiologists try this same maneuver, they are usually deeply frustrated: it has been estimated that 99 percent of the bacteria that can be found in the wild cannot be grown in the lab, in isolation from other bacteria.[15] Bacterial species, it turns out, are extremely interdependent: separate them from their bacterial symbionts, and they perish.

We have seen that macroorganisms rely on microorganisms for their well-being: recall our discussion of the role your gut bacteria play in your life. The converse is also true, though: microorganisms can rely on macroorganisms for their well-being. This is certainly true of the bacteria in your gut. It is also true, in a more surprising manner, of the tiny phytoplankton that live near the ocean's surface, where they can take advantage of sunlight. Whales feed deep down in the ocean, and when they surface to breathe, they defecate. The resulting fecal plumes—sometimes called *poonamis*—provide phytoplankton with the iron they need to flourish.[16] Consequently, eliminate the whales, and the phytoplankton are in trouble.

Like other living things, we humans are biologically dependent: deprive us of our fellow living things, and we will perish. We need to keep this in mind if we someday decide to leave Earth to travel to and then inhabit another planet. For us to survive, our spacecraft will have to carry a panoply of living organisms, including not only plants for us to eat and to supply us with oxygen, but all the organisms of the ecosystems those plants require in order to flourish. Our spacecraft would end up resembling Noah's ark. When all is said and done, it would probably be much more sensible to simply stay on Earth, curb our reproductive ardor, and take good care of our planetary home, than to abandon it in favor of some very distant "promised land."

BESIDES BEING INTERCONNECTED WITH EACH OTHER, the earth's living things are very much interconnected with its inanimate matter. If life had not arisen on Earth, its land, oceans, and atmosphere would be profoundly different than they are; and had the planet not changed the way it did, the earth could not support the living things it now does. In other words, not only do the earth's living things evolve, but they and the surface they inhabit can be said to coevolve.

Consider, to begin with, the chemistry of the earth's oceans. Two billion years ago, the oceans were full of dissolved iron that would have given them a greenish hue. The evolution of photosynthetic organisms, though, produced an abundance of O_2. The ocean's dissolved iron rusted in response to the presence of this oxygen, precipitated out of the water, and settled to the bottom of the ocean. The removal of this dissolved iron in turn made the earth's oceans habitable to a great number of species.

Some of the photosynthetic oxygen, rather than rusting iron, passed into the atmosphere where it again affected the evolution of life; indeed, land-dwelling animals like ourselves wouldn't be here without it. And of course, we humans have in turn had a profound impact on the earth's land, oceans, and atmosphere, as well as on the life forms that inhabit our planet.

Besides adding oxygen to the earth's atmosphere, living things have added water. By means of the process known as *transpiration*, plants draw water from the ground and release it into the air, in the form of water vapor molecules. The amount of water pumped in this manner is remarkable. A single tomato plant can transpire 34 gallons of water in a growing season, and a single corn plant can transpire 54. A single almond obviously doesn't contain much water, but to produce that almond, an almond tree has to transpire a gallon of water. Were it not for plants, all this water would remain in the soil or would sink to become part of deep aquifers.

Once airborne, water vapor molecules will typically remain independent of each other, but if conditions are right, they can combine to form water droplets. For this to happen requires the presence of a condensation nucleus. It can be a particle of dust or soot, or a salt crystal from ocean spray, but as we have seen, it can also be an airborne microbe. And besides themselves acting as nuclei, living things manufacture molecules that can play this role. Among them are the sulfate aerosols produced by phytoplankton[17] and the molecules of alpha-pinene that give pine trees their distinctive scent.[18]

The earth's oceans and atmosphere aren't the only things that have been affected by the presence of life; its minerals have as well. Were it not for life, there would be no limestone, since it is comprised of the shells of deceased marine animals, and no coal, since it is composed of decayed plant matter.[19] Life's impact on minerals is far greater than this, though. According to geologist Robert Hazen, 3,500 of the earth's 5,000 known minerals, with their distinctive chemical compositions, could not have formed if life hadn't arisen on Earth.[20] For one thing, without life, there wouldn't be as much atmospheric oxygen, and without that oxygen, a number of mineral-forming chemical processes would have been impossible.

We have already encountered Lynn Margulis, the biologist who proposed and then worked to gain general acceptance of the Big Gulp theory. Besides being an advocate of endosymbiosis, Margulis was an advocate of the Gaia hypothesis. She did not formulate this hypothesis—that distinction belongs to James Lovelock—but she did subsequently become one of its most important supporters. The Gaia hypothesis is concerned with the profound interconnection between the earth and the things that live on it. According to advocates of Gaia, it is a mistake to think of the earth and the life that inhabits it as existing independently; rather, they should be thought of as components of a single self-regulating system.

Different people interpret this hypothesis in different ways. New Age enthusiasts, for example, take Gaia to be the Mother Earth of a kind of pagan religion and attribute a form of consciousness to her, along with the ability to form and accomplish goals. Gaia, they believe, protects us and therefore deserves our protection. Scientific advocates of the Gaia hypothesis reject such claims as being "teleological," inasmuch as they attribute goal-driven behavior to inanimate objects. (In similar fashion, they reject the claim that evolutionary processes have a goal.) These same scientists nevertheless appreciate the way life affects the planet and the way the planet in turn affects life, and therefore believe that a complete, correct description of our planet will take the earth's surface and the organisms that inhabit it to be components of a living system.

I WILL CONCLUDE THIS CHAPTER BY DISCUSSING one other way in which Earth's life has interacted with its land, oceans, and atmosphere. We humans have had a dramatic impact on our planet. Because of us, the atmosphere has much more carbon dioxide than it otherwise would, along with much more smog. Because of us, the oceans have been polluted with oil, trash, chemicals, and sewage. It has also become more acidic as a result of the increased carbon dioxide in the atmosphere. And finally, besides changing the contours of the earth's surface, we have added a number of materials to it.

We humans are responsible for bringing into existence any number of man-made substances, including concrete, asphalt, and plastics. In 100,000 years, our descendants might dig a hole in the side of a hill, only to encounter a deposit of ancient plastic, the way we twenty-first-century humans might encounter a deposit of coal. If our descendants know their chemistry, they will conclude that in much the same way as a coal deposit is evidence for the past existence of forests on Earth, a plastics deposit is evidence for the past existence of clever but environmentally irresponsible beings—namely, ourselves.

Some might respond to this last comment by suggesting that it is optimistic to think that 100,000 years from now, our descendants will still roam the earth. Perhaps because of the changes we have made in the earth's land, oceans, and atmosphere, we will be extinct. I, for one, think such extinction is unlikely. In the 200,000 years we have existed, we humans have shown ourselves to be a remarkably adaptable species. Since emerging from the Rift Valley of Africa, we have gone on to inhabit radically different environments across the earth's surface. It seems likely that as long as future changes in the earth's land, oceans, and atmosphere are fairly slow—as they probably will be—we will survive.

It isn't clear, though, that there will still be room for 7 billion of us. Nor is it clear that a diminished population will be able to enjoy our standard of living. This might seem tragic, but realize that 2,000 years ago, the earth had only a few hundred million people on it, and those people lived what we would regard as primitive lives: there were no refrigerators, no toilets, and—get this—not even cellphones! Nevertheless, they seemed relatively satisfied with the lives they were living, as do we modern humans, and as our descendants a hundred thousand years from now likely will.

19

Your Many Afterlives

PEOPLE DIE ALL THE TIME. In fact, 150,000 will do so within the next 24 hours, meaning that 100 people will likely die while you read this page. What will become of them? And more to the point, what will become of you when you die?

How we answer this question depends on what we mean by *you*. The person that you are has, after all, both a physical component, your body, and a mental component, your mind. These components are obviously linked: what happens to your body affects what happens in your mind, and what happens in your mind affects what your body does. Nevertheless, most people take them to be separable. They think that after their body dies—and even after its component atoms have dispersed—their mind can continue to exist, meaning that they can remain conscious, continue to have ideas, and remember past events. Consequently, the question of what will become of you when you die can be divided into two questions: what will become of your body when you die, and what will become of your mind?

CHRISTIANS THINK THAT WHEN A PERSON DIES—when, that is, his body ceases to function—the body in question can come back to life and serve as the dwelling of his mind.[1] They might also spend time and energy practicing their faith in an effort to increase their chances of subsequent resurrection. If resurrection were possible, though, would you in fact want to be resurrected? Would you really want to take up residence in the body you had at the time of your death? Would you, in particular, want to spend eternity in a body that had died of severe burn injuries or been consumed by cancer? And if you weren't resurrected in your death body, what body would you occupy? Maybe your own body before it got injured or sick? But what if, as the result of a birth defect, your body had never been well?

According to St. Paul, your resurrected body, although the same body as the one you possessed at death, will be "glorified" by the resurrection process: it will have been transformed from a natural body into a "spiritual body." It isn't clear what this means, but we are told that the body in question will be immortal: it will never again experience death.[2]

Eastern religions believe in reincarnation rather than resurrection. On being reincarnated, your mind[3] will come back in a different body than the one you formerly inhabited. Your new body might be human, and if so, it might or might not be of the same sex as your former body. It is also possible that you will occupy not a human body but, say, that of a rat. The body transition process is, according to Theravada Buddhists, instantaneous. According to Tibetan Buddhists, though, complications can arise that delay the transition to a new body by as much as 49 days.[4] When your mind moves to its new body, it takes some of your memories with it. Consequently, you might be able to remember events from previous lives. The reliability of these memories is, of course, open to challenge.

At the emotional level, a belief in resurrection or reincarnation is perfectly understandable. People who enjoy life don't want it to end. Also, anyone who believes in "cosmic justice" will likely believe in either resurrection or reincarnation. There are good people who have miserable lives and bad people who don't, while alive, seem to pay any price for their misdeeds. By positing the existence of an afterlife, we can explain this state of affairs. Yes, evil people may have prospered while alive, but they will pay dearly for their misdeeds in their afterlife. They might, according to Christians, spend an eternity in hell, and might, according to Buddhists, come back as insects.

Meanwhile, the people they wronged while alive will be compensated in their afterlife. Christians, in particular, will get to spend an eternity in heaven. The cosmic scale of justice, then, can be thought of as a balance that has one pan in the hear-and-now and the other in the hereafter. And what will a heavenly existence be like? Heaven's inhabitants will get everything they desire and will therefore experience perfect happiness. Not only that, but they will experience it for an eternity. It is indeed a happy ending.

It is not at all clear, though, that the residents of heaven will end up any happier than they were on Earth. I say this because I am convinced that human happiness depends much more on the desires we form than on the circumstances in which we find ourselves. When we get the things we want, we might be satisfied for a time, but then we start taking those things for granted, form new desires, and are again dissatisfied. As a result of this process, we go through life unhappy, when happiness would be within our grasp, if only we could master our desires.[5] Suppose, then, that Christians, when they are resurrected, take their earthly personality to heaven with them, as well they should, since this personality is an important facet of their mind. Before long, their heavenly selves would get used to their new environment, start forming new desires, and again be dissatisfied—a dissatisfaction that would, in heaven, last for an eternity. Alas!

Buddhists, by way of contrast, understand full well the role desire plays in our happiness—that if we are unable to keep our desires within bounds, we will have a miserable life. The goal of Buddhists is therefore not to end up in a heaven in which we can continue satisfying our desires. The goal is instead to escape from the cycle of birth, death, and rebirth, so we can experience nirvana, in which the self is extinguished—*nirvana* means "blown out."

MODERN TECHNOLOGY HAS GIVEN RISE to other conceivable afterlives. Some people request that when they die, their body be frozen. Their hope is that someday, doctors will be able to cure whatever it was that killed them, meaning that they can be revived and continue their lives.

Some forms of cryonic suspension involve a kind of resurrection: the person's entire body is frozen, so that when he is brought back to life, he can continue to inhabit it. Other forms involve a kind of reincarnation: only the person's head is frozen, with the goal of subsequently grafting it onto a different body, or maybe implanting just its brain into a different body. Alternatively, the person's brain could be kept alive in a vat, with its nerves connected to a computer in a way that would make it possible for him to sense the outside world and communicate with other people.

The idea of a hybrid person, part brain and part computer, raises an interesting possibility: why not dispense with the brain altogether? Why not instead transfer the mind to a computer? Neurologists tell us that the brain is really just a very sophisticated analog computer. Interactions between neurons somehow give rise to the sensation of consciousness. Furthermore, our memories are somehow stored in these neurons. If we could figure out how neurons interact and if we could map the structure of the neurons in our brain, we could, in theory, create a computer model of it.

When this model was run, a mind—more precisely, *our* mind, with our ways of thinking and our memories—would come into existence. (Instead of a reincarnation, it would perhaps count as an e-incarnation?) This virtual mind could perceive the outside world through various sensors attached to the computer. Alternatively, the computer operators could fake sensory input in a way that would make us think we were inhabiting a world that didn't really exist, full of people who didn't really exist—or maybe with dungeons and dragons that didn't really exist. Done right, that virtual world would seem as real to us as our world does.

It will, to be sure, be exceedingly difficult for the brain ever to be mapped in sufficient detail that this model could be constructed—difficult, but not impossible. Indeed, ask a philosopher, and he might tell you that it is conceivable that at this very moment, your mind resides in a supercomputer somewhere, as does the virtual world that this virtual mind inhabits.[6] If this were the case, it wouldn't be hard for your virtual mind, on experiencing virtual death, to subsequently experience a virtual afterlife.

LET US NOW TURN OUR ATTENTION from the fate of your mind and its connection to your body, to the fate of your body. What, in particular, will be the fate of the atomic you?

Many people request that when they die, they be buried. If their wishes are respected, and if they are buried in a wooden casket, their body will subsequently decay, and before long, their casket will be empty. This, at least, was what Murray G. Motter discovered when, during the summers of 1896 and 1897, he disinterred 150 graves within the city limits of Washington, DC.[7] Motter could tell, by reading the tombstone, how long the person he was digging up had been buried. He found considerable variability in the manner in which corpses decayed and in how long it took them to do so:

> I have found the bones, after an interment of seventy-one years, still preserving their general form and appearance, though easily crushed between thumb and fingers; the hair I have seen practically intact after thirty-six years. The brain I have found a still recognizable grayish mass, lying within the skull after all the other soft tissues had disappeared and the skeleton had been completely disarticulated. Indeed, I have found it, after eighteen years and two months . . . , lying on the occipital bone after the skull itself had fallen apart.[8]

Motter kept track of what insects he found in coffins at the various stages of decay and thereby became a pioneer in the field known as *forensic anthropology*. As part of their investigations, modern forensic anthropologists expose donated bodies in various ways and monitor the decomposition process. They have found that bodies are thoroughly recycled by nature. Bacteria start playing a role within minutes after death. After a few hours, if a body is on the surface of the earth, insects will pay a visit. Flies will land on it and lay their eggs which, on hatching as larvae, will feast on the corpse's flesh. Carnivorous birds and animals might come to share the feast. If the body is buried, worms will find and devour it. The bones might be scattered by animals, but even if this happens, the place where the corpse decayed might be detectable by the luxuriant growth of vegetation there. The plants in question will be feeding on the matter that leaked from the corpse. This research is very useful to police who come across a corpse and need to estimate how long it has been dead.

Many people, having taken meticulous care of their body while alive, will be disturbed by the prospect of that body rotting, and if they believe in resurrection, they might want to keep their body in good shape for later use. They therefore take steps to prevent the posthumous decay of their body and the dispersion of its atoms. Consequently, rather than being buried in a wooden casket, they might insist that they be buried in one that is decay proof and hermetically sealed. And better still, instead of burying that coffin, they might insist that it be put in an above-ground mausoleum.

This will, to be sure, protect their corpse from being ravaged by things that live outside their body, but it neglects the billions of organisms that, as we have seen, live inside it. There are microbes in your microbiome that play a minor role while you are alive but flourish when you die, and thereby become what is known as your *necrobiome*. These microbes have the power to thoroughly rot your body without any assistance from the outside world.[9] If you want to protect your corpse against decay, then, you will have to protect it against both the external world and your own microbiome. This can be accomplished, to a degree, by having your body embalmed.

Although Vladimir Lenin had requested that he be buried when he died, the powers that be decided that it would be politically strategic to put his corpse on public display, and after nearly a century, that is where it remains, in a climate-controlled glass coffin. It looks like the body of someone who died only yesterday. This remarkable state of preservation is the result of initial embalming, followed by subsequent chemical interventions. There have also been physical interventions: as parts of him decayed, his caretakers resorted to cosmetic surgery, in which they replace bits and pieces of him. What we therefore end up with, in the case of Lenin, is another version of the ship of Theseus paradox, this time involving body parts.

The embalmed corpses of ancient Egyptians give us an indication of what becomes of embalmed bodies in the long run. They desiccate and change color. Although they remain identifiably human, they are a transmogrified form of their living selves. Could even their own mothers identify them? And significantly, these corpses weren't merely embalmed; they also had organs, including their brain, removed. They therefore wouldn't be ideal candidates for resurrection.

Although embalming may reduce the damage microbes do to a corpse, it will not prevent the process of autolysis, in which enzymes produced by your cells start digesting you. To stop the action of these enzymes, it helps to dehydrate or freeze the corpse. Five hundred years ago, the Incans sacrificed children high up in the Andes and buried their corpses there. In this cool, dry environment, these corpses have aged remarkably well. Ötzi the Iceman, mentioned back in chapter 2, was apparently covered with snow shortly after he died and subsequently remained snowbound until his body was discovered 5,000 years later. As a result, it is quite well preserved (see Figure 19.1).

Another way to preserve a body is to sink it in a peat bog, preferably in winter. The water will be cold, acidic, and anaerobic, conditions well suited to the body's preservation. As a result of being pickled in this manner, the body will turn dark brown, but facial features will be remarkably intact. This is certainly the case with Tollund man, whose remains were found in 1950 in a peat bog in Denmark. Even after 2,400 years, his mother would be able to identify him (see Figure 19.1).

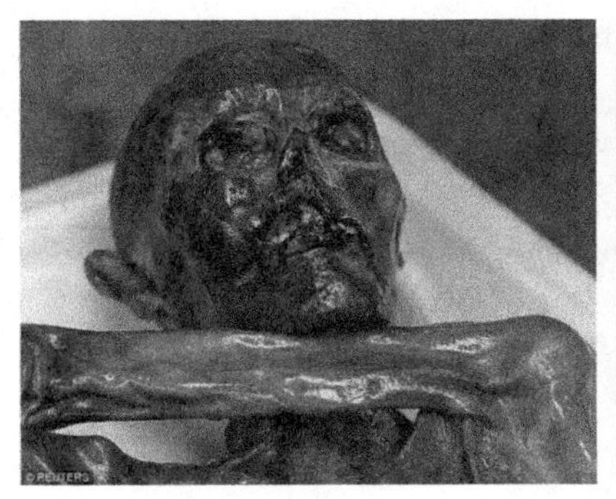
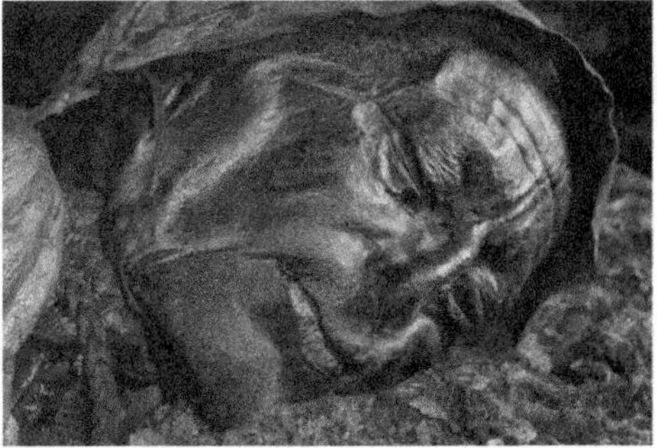

FIGURE 19.1. Resisting the ravages of time. On the left is the 5,000-year-old corpse of Ötzi the Iceman and on the right is the 2,400-year-old corpse of Tollund man.

SOME PEOPLE, RATHER THAN BEING REPELLED by the prospect of having their body rot and be consumed by living things, embrace it. Because it is hard to dig in the ground where they live, and because wood is in short supply, Tibetans might, instead of burying or cremating a corpse, resort to "sky burial" to dispose of their dead. They leave a naked corpse out in the open. Soon, vultures come to consume it. When the skeleton has been picked clean, human "body-breakers" encourage further consumption by cracking open the skull and other bones. Before long, only small bone fragments remain. The corpse will have been biologically recycled.

Many in the First World would find this a gruesome and disrespectful way to treat a body, but there are exceptions. Author Edward Abbey was one of them. He requested that when he died, his friends bury his unembalmed body out in the desert, where it would "fertilize the growth of a cactus or cliff rose or sagebrush or tree." He also requested a funeral at which, rather than solemn speeches, there would be "lots of singing, dancing, talking, hollering, laughing, and lovemaking," spurred on by "a flood of beer and booze!"[10]

Evolutionary biologist W. D. Hamilton also embraced the biological recycling of his body, but his plans along these lines were rather more ambitious than those of Abbey. He wanted his body transported to the forests of Brazil, where it would not be buried but would instead be laid out in a manner that, although possums and vultures would not be able to consume it, insects—and in particular, Coprophanaeus beetles—would have ready access to it. These beetles are the size of golf balls and have iridescent violet wings. Hamilton wanted them to find his body and lay their eggs on it. The resulting offspring would consume it, and when they had finished the task, would fly into the sunset en masse, buzzing like a swarm of motorbikes.[11] What a splendid way to return your atoms to the world!

I was later disappointed to learn that when Hamilton died in 2000, these instructions had not been carried out. His family instead buried him in the conventional manner in England. Hamilton's devoted companion Luisa Bozzi, with his wishes in mind, had a bench placed near the grave. On that bench the following words are inscribed:

> BILL. Now your body is lying in the Wytham Woods, but from here you will reach again your beloved forests. You will live not only in a beetle, but in billions of spores of fungi and algae brought by the wind higher up into the troposphere, all of you will form the clouds and wandering across the oceans, will fall down and fly up again and again, till eventually a drop of rain will join you to the water of the flooded forest of the Amazon.[12]

Although it is likely that many of Hamilton's atoms will someday reach the Amazon, the trip would have been shortened if he had been cremated rather than buried. The atoms of a buried body will ultimately be transported to the surface for release into the atmosphere, but it can take years or decades for this to happen. When a body is instead cremated, it happens in a few hours. The water inside the body turns into steam and escapes through the crematorium chimney as water vapor molecules. When the fats, sugars, and proteins in the body are consumed in the flames, their carbon and hydrogen atoms merge with atmospheric oxygen to become part of carbon dioxide and water vapor molecules, which also go up the chimney.

When the flames die down, most of the cremated body's atoms will be gone. Only a few percent of them, including the calcium and phosphorus atoms in the bones, remain as ash. How much is left behind depends on how much a person's bones weigh and their density. A typical adult will produce between 3 and 9 pounds (1.4–4 kilograms) of ash, with men producing more than women. This ash might be said to represent a person's earthly remains, with the water vapor and carbon dioxide molecules described earlier being his atmospheric remains.

It is entirely possible—highly likely, in fact—that these latter remains will become part of another living thing. One of the water vapor molecules, for example, might become part of a snowflake that a schoolchild catches on her tongue. One of the carbon atoms might, as part of a carbon dioxide molecule, be taken in by a plant. It might thereby take up residence on someone's lawn or as part of a giant oak tree. Alternatively, it might become part of a corn plant, the corn of which is eaten by a mouse, which in turn is eaten by a hawk. If this is its fate, the carbon atom will get to spend a few months of its billions of years of existence soaring over fields and prairies. And if the corn, instead of being eaten by a mouse, is eaten by a cow, which in turn is eaten by a human, the carbon atom will take up residence in that person. He might be a famous actor or a brilliant musician, or maybe a serial killer. In any case, the atom will get to enjoy a kind of atomic reincarnation.

The fate of the cremated person's ashes is up to his descendants. These earthly remains might end up in a box in a closet or in an urn on a mantelpiece. They might also be returned to the environment, maybe on a beach the person loved or in the stream in which he once caught salmon. In such cases, there is a good chance that the atoms in the ash will again become part of a living thing. A calcium atom, for example, might take up residence in the bones or shell of an animal.

On various occasions, when relatives have asked what I want to have happen to my body when I die, I have told them that I want to be cremated, and I want my ashes to be sprinkled in a garden, so that some of my atoms can subsequently be served up, on a fine summer day, as components of a tomato salad. They laugh and assume that I am joking, but I don't think I am.

SO MUCH FOR THE FATE OF THE ATOMIC YOU. When you die, your atoms will, unless you take extreme measures, return to the environment, where many of them will again become part of a living thing. Let us now turn our attention to the fate of the cellular you. You might assume that when you die, your cells die as well, and this for the most part is true, but there are some interesting exceptions.

To begin with, your cells can outlive you if you donate an organ—say a kidney. It will be only a few percent of your cells, but they can live on for decades. Your cells can also outlive you if they become cancerous. This is what happened to Henrietta Lacks, who died in 1951. A researcher cultured some of her cancer cells. The descendants of these cells are not only still living but are thriving in laboratories around the world. It would appear that the cancer that caused most of Lacks's cells to die conferred cellular immortality on some of them.

Those who are lucky enough to avoid cancer can attain a kind of cellular immortality by having children. When your gamete fuses with that of your mate, it does not die. It continues to live, albeit in merged form, first as part of a zygote and then as part of the cells that descend from that zygote. This means that for as long as your offspring keep having offspring, the cellular you will in some sense continue to exist—especially if you are a mother.[13]

THIS BRINGS US, FINALLY, TO THE GENETIC YOU. Suppose you wanted to achieve "genetic immortality"—more precisely, you wanted your genes to survive your death and, ideally, to be around for many more generations. The obvious way to accomplish this is by having children. You would thereby pass on your genes—actually, only half of them—to each child you had.

As we saw in chapter 17, though, you don't need to worry about the survival of your genes because they aren't really "your" genes; they are instead copies of genes that exist independently from you. Because other people and even members of other species share your genes, they will survive your death, even if you die childless.

On realizing this, you might shift your genetic goal. What is important, you might conclude, is not that "your" genes survive your death, but that your genome does. What you want, in other words, is for humanity to possess the exact genetic recipe for you in case they ever felt a need to bring you back. You could accomplish this by having your DNA preserved or by having it sequenced and then carefully storing the resulting data.[14]

Taking such steps would confer potential genetic immortality on you, but what if humanity never felt it necessary to resurrect the genetic you? Even worse, suppose someone carelessly threw away your DNA sample or deleted your genome from a database? The genetic you would then be lost forever.

With such fears in mind, you might take the ultimate genetic step and have yourself cloned.[15] Your genome would then enjoy at least one more generation of embodiment, and your genomic resurrection would have been achieved. Not only that, but if your clone felt similarly protective of his or her genome, the process could continue for very many generations. This would probably be as close to genomic immortality as one could hope for.

What good, though, would this sort of genetic afterlife do you? The resulting person would no more be you than an identical twin, if you have one, is you. It would lack your memories, your way of thinking, and much of your personality. Preserving your genome might gratify your ego, and might benefit future geneticists doing research on the history of the human genome or doing comparative genomic studies, but it will not in any meaningful sense provide you with an afterlife.

THIS CONCLUDES OUR DISCUSSION OF YOUR AFTERLIFE. Your atoms will certainly survive your death, as might some of your cells. Your genes will likewise survive your death, although your genome probably won't. Furthermore, it might be possible, with technological or divine intervention, for your body to survive your death. As survivals go, however, the survival of your atoms, cells, genes, or even your body are not nearly as important as the survival of your mind. It is, after all, the component of you that plays the biggest role in making you the being that you are.

As we have seen, though, it is not at all clear that there will be a mental life after death, and if so, what form it will take. Maybe your mind will go to heaven—an experience which, as we have seen, might or might not be to your liking. Or maybe your mind will simply continue to dwell in a computer simulation. Be this as it may, one thing is certain: before you experience a life after death, if there is one, you will experience a life before death. This in turn suggests that you would be wise to spend your time and energy living this life as well as you can. It is with this thought in mind that we turn our attention, in the final chapter of the book, to the question of the meaning of life.

20

Why Are You Here?

IN THIS BOOK, WE HAVE EXPLORED your deep history as well as your future, seen from different points of view. If I have done my job as scientific historian, you now have a much better understanding of how you came to exist, as well as how remarkably contingent your existence is. Think of all the things that would have prevented you from reading these words: if you had been on the top floor of the North Tower of the World Trade Center at 8:45 am on September 11, 2001; if a different sperm had made it to the egg that became you; if your parents hadn't met; if your grandparents hadn't met, and so on, back into your ancestry; if phytoplankton hadn't oxygenated the earth; if Lynn, the first eukaryote, hadn't come into existence as the result of a chance encounter between an archaeon and a bacterium; if life hadn't arisen; if the moon hadn't come into existence as the result of a collision between a planet and the earth; and if one or more supernovas hadn't created the debris field in which the sun and Earth formed. Seriously, you are lucky to be alive!

To this list, of course, we must add one other significant event: the Big Bang. If it hadn't happened, you wouldn't exist for the simple reason that nothing would exist. Although scientists have provided an incredibly detailed description of what has happened since the Big Bang, they find it much harder to explain the event itself. They can't, in other words, answer a very basic question: why is there something rather than nothing? And if they could answer it, their answer would immediately give rise to another very basic question: given that there *is* something, why *this* something rather than *some other* something?

Along these lines, consider the dimensionality of our universe. Space—or rather space-time—has four dimensions, three of which are spatial and one of which is temporal. Why four? Why not only two spatial dimensions? For that matter, why not a hundred? And these aren't the only why-not questions that arise. Our universe is governed by laws of nature. For instance, mass and energy are connected in accordance with the famous law $E = mc^2$, where E is energy, m is mass, and c is the speed of light. But why this law rather than some other law? And why does *c*, the speed of light, have the value that it does, *exactly* 299,792,458 meters per second?[1] Why not 299,792,452.97 meters per second? For that matter, why not a leisurely 12 meters per second?

The values of the constants in the laws of nature are quite important. In most cases, if those values were slightly different, we wouldn't be here. For example, if the gravitational constant, which is equal to 0.0000000000667408 $m^3\ kg^{-1}\ s^{-2}$, were slightly lower, gas clouds would have had to be much bigger before they "ignited" to become stars. The resulting oversized stars would burn through their fuel before life could evolve in the planetary system of that star. Conversely, if the gravitational constant were slightly higher, gas clouds would ignite when they were still small, meaning that the resulting stars would be smaller, wouldn't produce a lot of light, and would quickly run out of fuel.[2] Again, life would have a hard time arising in the planetary system of such a star. And before we move on, one more comment is in order. There is a chance that the physical constants are not constant, that they are instead changing with the passage of time. Such changes, by the way, could conceivably threaten our continued existence.

MOST CULTURES, CONFRONTED WITH THE PROBLEM of explaining the existence of the universe, have come up with creation stories. The Tungusic peoples of Siberia, for example, think that in the beginning there was a primordial ocean, along with a deity named Buga who set fire to it to create land. In Hinduism we are told that "In the beginning rose Hiranyagarbha, born Only Lord of all created beings. He fixed and holdeth up this earth and heaven."[3] In the creation story favored by Jews, Christians, and Muslims, God created the heavens and the earth. These creation stories, however, don't get to the bottom of things, inasmuch as they raise new questions. In the case of the Tungusic story, we can ask who created Buga, as well as who created the primordial ocean; in Hinduism, we can ask how Hiranyagarbha arose; and in the case of Judaism, Christianity, and Islam, we can ask who created God.

In the eleventh century, St. Anselm came up with a clever answer to this last question: no one created God. No one had to, since God doesn't just happen to exist, the way you do; he instead exists *necessarily*, meaning that it is impossible for him *not* to exist. Anselm's defense of this claim begins by characterizing God as *the being than which no greater being can be imagined*. A being that exists is obviously greater than one that doesn't exist, and a being that necessarily exists is greater than one that just happens to exist. Therefore, given who he is, God necessarily exists.

This line of reasoning, known as the *ontological argument*, has been challenged,[4] but for the sake of this discussion, let us assume that its conclusion is correct—that God necessarily exists. We are then left with the question of why he created this universe rather than some other. And if you answer that he created this universe because it is the one best suited to the existence of humans, we are left with the question of why he felt compelled to create us.

For insight into why God created this universe, with ourselves as its inhabitants, many will turn to the opening pages of the Book of Genesis in the Old Testament. There we are told that on the sixth day, he created mankind, both male and female, in his image and instructed them to "be fruitful and increase in number; fill the earth and subdue it." He went on to give them dominion over all other living things. This implies that God created the universe for our benefit—even that he created it in order to create us, an implication that should make us feel very special.

If we turn to the second chapter of Genesis, though, God's intentions become unclear. We are told that he created not mankind, male and female, but Adam alone, and the implication is that God created him because he needed a gardener.[5] He subsequently went on to create Eve, but only as an afterthought, on realizing that Adam was lonely. Eve was created, however, not as a reproductive partner for Adam, but merely as his helper. The evidence for this is that Adam and Eve initially were psychologically asexual beings who were not aroused by each other's nakedness. For them to become interested in sex, they had to partake of the fruit of the tree of the knowledge of good and evil, something God had specifically forbidden them to do. More precisely, God warned Adam that if he ate from this tree, he would "certainly die"—a threat, by the way, that God didn't follow through on. It was only after they had eaten the fruit that "the eyes of both of them were opened, and they realized they were naked." They experienced, in other words, sexual desire.

This suggests that God did not intend to bring humanity, including ourselves, into existence. Instead, he intended to create only one male, and when he realized that this wasn't working, added one female. It was only because the male and female in question disobeyed his commands that we are here. If anything, God, rather than loving us, should be annoyed by our existence. We are, after all, living evidence of his past blunders. Then again, as an infinitely wise being, he should have seen it coming.

More generally, why would God want to bring humans into existence? If it was so we could worship him, why would an infinite being care about being worshiped by pathetic little creatures like ourselves? If it was instead because he loved us, why doesn't he take better care of us? Why, for example, does he allow millions of us to starve to death when he can, simply by willing that it be so, make manna fall from the heavens to feed us? He did it for the followers of Moses.[6] Why not do it again? Why not perform yet other miracles to prevent innocent people from being drowned by tsunamis and killed by earthquakes and tornadoes? And while he is at it, why not miraculously cure all cases of cancer? For an infinite God, it takes only infinitesimal effort to perform miracles. If God loved us, wouldn't he perform many more of them on our behalf?

And one other thing: if God created the universe as a dwelling place for us—or rather, for Adam—it is a gift that can only be described as mind-bogglingly lavish and wasteful. Surely he could have created the earth without having placed it in an infinite universe. In particular, he could have put the earth inside a finite dome to which the stars were attached—the way Genesis says he did. What he instead has done is the cosmic equivalent of a father building a trillion-room mansion for his son, with only one of those rooms being habitable. What a waste!

SCIENTISTS HAVE PROVIDED US WITH THEIR OWN "creation story" in the form of the Big Bang theory. It describes in remarkable detail what must have happened for our universe to be the way it is, and it takes us all the way back to within a split second after the Big Bang event. At that point, though—just when we think the mystery is going to be solved—the theory goes silent, leaving us with two very fundamental questions: why was there a Big Bang event, and given that there was, why did it result in this universe rather than some other?

Scientists who favor *multiverse theory*, as it is known, think they have an answer to these questions. They propose that ours is just one of many universes that exist; indeed, some assert that every universe that possibly could exist does in fact exist, meaning that ours is one of infinitely many universes. Although most of these universes, because of the laws of nature they conform to, will not harbor intelligent life, ours is an exception. There are likely others as well.

This is an interesting theory to entertain, as well as an entertaining theory, as employed by science fiction writers. As science, though, the theory faces a serious challenge: how can it ever be proved or disproved, given that we don't have access to these alternate universes? And more to the point of the current discussion, multiverse theory raises a new why-question: even if we agree that our universe exists because it is part of a collection of universes, why does the collection exist? Who or what created it? And if nothing created it, how and why did it spontaneously come into existence?

A philosopher might respond to our attempts to "reach bottom" in our exploration of the universe by asserting that the universe we are trying to explain doesn't in fact exist. It is instead an illusion. It might be a dream we are having. Or maybe we are just characters in a computer simulation, meaning that we and our seeming universe are only virtually real.[7] As a philosopher, I am perfectly willing to admit that I might be dreaming or even that I might be a character in a computer simulation. Having made this admission, though, I hasten to add that this possibility, even if true, doesn't get to the bottom of things. Dreaming requires a brain, leaving us with the question of why my brain exists. Likewise, computer simulations require computers and programmers, leaving us again with the question of why they exist.

At this point, a philosopher might completely eliminate the physical world from his explanation and instead assert that I exist as a mind that is capable of thought in the absence of all physical things, including brains and computers. But of course, this still doesn't get to the bottom of things, inasmuch as my inquiring mind will want to know why it exists.

THE PROBLEM WITH TRYING TO GET TO THE BOTTOM OF THINGS by answering why-questions is that our answers generate new why-questions. Tell me that God created everything, and I will ask why he exists, why he created the universe, and why he created this universe rather than some other universe. Tell me that the Big Bang was the beginning of everything, and I will ask why it happened and why it resulted in this universe rather than some other. It becomes clear that to truly get to the bottom of things, we will have to answer why-questions in a way that does not give rise to new why-questions, but is this possible?

Children of a certain age discover the power of why-questions. They can respond to almost anything an adult says by saying "Why?" and the adult will talk some more. When their question has been answered, they can again ask "Why?" and again the adult will talk. In the process of asking these questions, children learn a lot about the world. They also discover, much to their astonishment, that most adults are only a few why-questions away from having to admit their ignorance about the world:

CHILD: "Why do I have to eat broccoli?"
PARENT: "Because it is good for you."
CHILD: "Why is it good for me?"
PARENT: "Because it has vitamins."
CHILD: "Why are vitamins good for me?"

At this point, most parents get flustered, for the simple reason that they don't know what, exactly, vitamins do and therefore don't know why we need them. Children will also discover that most adults can't even tell you the most basic of things, such as why the sky is blue.[8]

Suppose that in our effort to get to the bottom of things, we figure out how to answer basic why-questions without thereby giving rise to new why-questions. Our feeling of success will likely be short lived, inasmuch as it won't be long before some smart-aleck philosopher drops the bomb: "Why is it that *these* answers to why-questions don't give rise to new why-questions, the way the previous answers did? What makes them different?" Sigh!

Don't get me wrong. If we want to learn more about the world, we should ask why-questions, lots of them. But in doing so, we need to keep in mind their recursive nature: why-questions beget more why-questions. There comes a point, though, at which asking why-questions, rather than adding to our understanding of the universe, becomes little more than a diversion from the pursuit of useful knowledge.

AS A RESULT OF DOING THE RESEARCH FOR THIS BOOK, I have come to the conclusion that we live in a universe that is "bottomless," in the sense that it will never be possible to reach the end of the chain of why-questions we can ask about our existence. There will always remain an element of mystery. This is an admission, I realize, that will leave some readers unsatisfied. They want very much for their universe to have a bottom.

This desire is completely understandable, and yet I suspect that it is misguided. One of the reasons people want the universe to have a bottom is that they fear that otherwise, there is no reason for their existence, and if this is the case, how can their life have any meaning? This same kind of thinking causes many people to turn to religion. They want to hear that God has a plan for them. Such a plan, they feel, gives meaning to their life. But it isn't at all clear how someone else's having a plan for my life can make it meaningful. To the contrary, I think that the first step to take, in order to have a meaningful life, is to consciously form a plan for your life and then put that plan into action. The life you subsequently live may not have cosmic significance, but that is what you would expect in a cosmos without meaning. It will nevertheless be personally significant, which is the most you can hope for in a bottomless universe.

As a philosopher, I have thought, taught, and written about meaning-of-life questions. I have come to the conclusion that most such questions are fundamentally misguided. To better understand this claim, suppose someone came up to you, showed you a pencil, and asked what it means. In an attempt to clarify this rather strange question, you might inquire whether the person was asking what the word *pencil* means. Suppose he replied that, no, he was asking what *the pencil itself* means.

My answer would be that the pencil itself has no meaning, in this sense of the word *meaning*. I would add that in asking this question, the questioner is committing what philosophers call a *category mistake*: he is asking whether something has a quality that only something in a different category can have. If you ask me for the diameter of hope or the location of the number six, you are making a similar error: desires cannot have physical dimensions, and numbers are not located in physical space—although the numerals used to designate them may be.

I would add that although a pencil cannot have meaning, it can be used to do meaningful things, such as make grocery lists or solve a sudoku puzzle. In much the same way, although your life may have no meaning in and of itself, you can use it to do meaningful things. In particular, you can confer meaning on it by living in accordance with a plan you have devised for your life.

BEFORE I BRING THIS BOOK TO AN END, let me make one last comment about life in a bottomless universe. I am of two minds about it. On the one hand, I would like very much to get to the bottom of things—to know why our universe exists and why it is the place it is. On the other hand, I suspect that I would be saddened if I ever did get to the bottom of things. It would mean the end of why-questions about the universe and therefore the end of the little burst of delight that comes when, after some research, I find out why, according to science, things are the way they are.

Furthermore, the realization that I don't know why *all this* exists, rather than causing me to slip into depression, has transformed me into a very appreciative person. When I wake up in the morning and see that my glasses are still on the nightstand where I left them, I conclude that the universe is still there, with the same physical laws and constants. I know full well that I can't explain why this should be so. Nor can anyone else. It is a mystery that not only is unsolved but likely never will be. At the same time, it is a mystery that, as a result of trying to solve it, has had the effect of making me profoundly grateful to still be a part of this universe. I get another day of life, another chance to get things right! And if all goes well, I will be able to spend part of that day asking why-questions. If there is a heaven, it can't be much better than this.

We inhabit a fundamentally mysterious universe. The most sensible psychological strategy, in such a place, is to embrace its mystery. And if you want to fully appreciate the life you find yourself living, it makes sense for you, at random moments, to pause in whatever you are doing to reflect on the remarkable series of events, described in the preceding pages, that made your life possible. Lucky you!

NOTES

INTRODUCTION

1. The three domains of living things are eukaryotes, bacteria, and archaea. You are a eukaryote; the microorganisms that live on and in you are, for the most part, bacteria and archaea.

CHAPTER 1

1. Balaresque 2015.
2. "Names" n.d.
3. "Not Smith and Jones" 2011.
4. How many years an average generation lasts is open to debate. See "Generation Length" (2015) for a useful discussion of the issues involved. I am using 25 years as my average to keep the math simple.
5. The top line of the tree would be 1.2×10^{24} centimeters = 1.2×10^{19} kilometers wide. The diameter of the solar system—understood to be the diameter of Neptune's orbit—is 9×10^{9} kilometers. Notice that $(1.2 \times 10^{19})/(9 \times 10^{9}) = 1.3 \times 10^{9}$.
6. The top line of the tree would have 10^{24} spaces × 10 bytes/space = 10^{25} bytes of data. A terabyte computer can store 10^{12} bytes of data. Notice that $10^{25}/10^{12} = 10^{13}$, which is 10 trillion.
7. The age of our species is the subject of paleoanthropological debate, with some people arguing for a substantially earlier "birth." See Hublin 2017.
8. The date of the migration from Africa is disputed. Furthermore, it is unlikely that there was a single migration; rather, there were multiple migratory pulses, some of which might have taken place before 70,000 BC. See Parton 2015 and Gibbons 2015b. And realize that while some people might have migrated from Africa, others would have migrated there. See Llorente 2015.

9. Pugach 2013.

10. Archaeologists refer to any member of the genus *Homo* as a *human*. This includes members of the species *Homo sapiens*, which are referred to as *modern* humans, as well as *archaic* humans, including members of the species *Homo neanderthalenis, Homo heidelbergensis*, and *Homo erectus*. The first human species was probably *Homo habilis*, which evolved 2.8 million years ago from the non-human genus Australopithecus.

11. Genesis 12:1–5.

12. Balter 2014.

13. Gibbons 2015a.

14. The question of when we parted ways, evolutionarily speaking, with chimpanzees is much debated. See Curnoe 2016 for a discussion of this debate. I will not attempt to settle the matter in these pages.

CHAPTER 2

1. Lanier 2000.

2. Martha Jefferson's father John Wayles was a planter and slave trader. He had Martha with his first wife, Martha Eppes. She subsequently died, as did two more wives. At that point, Wayles apparently began a relationship with Betty Hemings, Sally's mother. Wayles is therefore the father of both Jefferson's wife and Sally Hemings, making them half-sisters.

3. "The Time I Accidentally Married My Cousin" 2013.

4. Barton 2008.

5. Carter 2012.

6. Haub 2011.

7. "New App Urges Icelanders" 2013.

8. Elhaik 2014.

9. Conger 2012.

10. Cassidy 2016.

11. Durant 1963, 452.

12. Wilkinson 2008.

13. Connor 2008.

14. "Nomenclature of Inbred Mice" n.d.

15. Main 2014.

16. Kolbe 2012.

17. Lizards can be pregnant in the sense that they can store the sperm necessary to fertilize their eggs. For more on anole lizard mating habits, see Walls n.d.

18. Hein 2004.

19. To put this number into perspective, realize that if you and your spouse have two children, you will each be responsible for bringing 1.0 persons into existence; if you have three, you will each be responsible for bringing 1.5 into existence.

20. Notice that $1.28^{92} = 7.3$ billion.

CHAPTER 3

1. Whereas in a human being, the brain might represent 2 percent of body weight, in a small ant, it might represent 15 percent. See Seid 2011.

2. Fields 2008.

3. Marino 2007.

4. Salvini-Plawen 1977. This assertion has subsequently been challenged.

5. Although birds are descendants of dinosaurs, they are not descendants of pterosaurs, meaning that they discovered flight independently.

6. For a discussion of bipedalism in animals, see Alexander 2004.

7. Part of this uprightness is an illusion. Penguins' femurs are parallel to the ground when they walk; it is only their lower legs that are perpendicular to it. See the penguin diagram in Thomas 2015.

8. Wichura 2015.

9. Sockol 2007.

10. Zihlman 2015.

11. Zihlman 2015.

12. Rogers 2004.

13. Zihlman 2015.

14. Tattersall 2015, 66.

15. Tudge 1996, 256.

16. Roach 2013, 483.

17. Perkins 2013.

18. Young 2003, 166.

19. Young 2003, 170.

20. We are not the only animals that kill by throwing rocks. Egyptian vultures use their beaks to throw rocks to crack open ostrich eggs to consume their contents. By doing so, they kill the developing ostrich within.

21. Pobiner 2016.

22. Roach 2013, 483.

23. The dates and order-of-invention of the atlatl and bow and arrow are uncertain because the wooden components of these weapons are unlikely to survive the ravages of time.

24. Chatterjee 2015.

25. I subsequently discovered that I had not been alone in doing such research. See Young 2003.

26. Young 2003, 170.

27. Morgan 2013.

28. To be sure, language isn't essential in cooperative hunting, as the activities of lions and hyenas demonstrate. More sophisticated forms of cooperation, however, are impossible without language.

29. Horan 2005.

30. Wrangham 2009, 139–140.

31. Wrangham 2009, 97.

32. The food itself would have been more nutritionally dense, and because it was cooked, it would have required less energy for your digestive system to gain access to that nutrition. For a discussion of the impact cooking has on the caloric content of food, see Twilley 2016.

CHAPTER 4

1. Mora 2011 estimates that there are 8.7 million eukaryotic species. To these, we must add prokaryotic species, which are certain to number in the millions. I am using 10 million as a nice, round, lower-bound number.

2. The percentage of species that have gone extinct is deduced by comparing the number of species that currently exist with the number of species that have ever existed. Both numbers are highly speculative, meaning that the resulting percentage is also speculative, but the 99-percent estimate offered by many sources seems plausible.

3. Tattersall 2015.

4. More precisely, the asteroid impact killed the bigger dinosaurs. Some of the smaller ones survived the event and gave rise to modern birds. There is reason, by the way, to think that multiple factors contributed to the extinction of the dinosaurs, with the asteroid strike perhaps representing the "last straw." There is also reason to think that the dinosaurs were in decline long before 66 million years ago. See Sakamoto 2016.

5. The proper term for what tenrecs do is not *hibernation* but *estivation*, since they go dormant not in the winter but during hot, dry seasons.

6. White 2002.

7. Schrag 2002.

8. As we shall see, this generalization is "almost true." And while I am generalizing, here is another: almost every generalization we try to make regarding living things will have an exception. This is what one would expect, though, given that life arose on Earth not as the result of the implementation of a grand plan but as the result of countless jury-rigged solutions to the problems that organisms encountered.

9. In particular, species members that died before having offspring would not appear on your family tree.

10. Frazer 2015.

11. Kuban n.d.

12. Gordon 2014.

CHAPTER 5

1. Other, even more complex combinations of chromosomes, such as XXY, are possible. Individuals with this genetic makeup might self-identify as male, female, or otherwise, and they might or might not be fertile.

2. The fossil record indicates that by 1.2 billion years ago, sexual reproduction was taking place. See Butterfield 2000.

3. Many factors play a role in our sexual preferences, including our genes, our brain structure, and our environment, both in the womb and during our childhood.

4. For an examination of our wiring and the role it plays in our life, see Irvine 2006.

5. Calling them *daughter cells*, as biologists commonly do, implies that they are female, but this is not the case. They could just as well have been called *son cells*. And since they are asexual, it would make more sense to refer to them simply as *cellular offspring*.

6. Kindlmann 1989.

7. Hales 2002.

8. Lane 2009, 123.

9. For a discussion of this debate, see Scudellari 2014.

10. Kirschner 2005, 94.

11. Ross 1978. For a broader discussion of the phenomenon of sex changes in fish, see "Sex Change in Fish Found Common" 1984.

12. Lane 2006, 235.

13. Quirk 2013.

14. Lane 2006, 236–237.

15. Jacob 1977. The suggestion is not that evolution is capable of intent, the way a human tinkerer would be; evolution is mindless. The suggestion is instead that the end result of the evolutionary process resembles the end result of mindful tinkering.

CHAPTER 6

1. Green 2010.

2. Recall that when one individual appears on a family tree, so do all the ancestors of that individual. Consequently, if we assume that Neanderthals will have Neanderthals as ancestors, it will be impossible for there to be only one Neanderthal on your family tree.

3. This is what is known as the *biological species concept*; it is only one of many species concepts currently in use. See de Queiroz 2007, 880.

4. This, to be sure, would be an impossible task for one person to complete—indeed, even a talented team consisting of thousands of time-traveling scientists would find it daunting—but I will not let this interfere with my storytelling.

5. It is not clear that time travel is possible. For an introduction to the kinds of paradoxes that arise in conjunction with time travel, see Christoforou 2014.

6. Quinn 2013.

7. De Graciansky 2011, 359.

8. Lyons 2014.

9. Surridge 2003.

10. The dinosaur ancestors of modern birds would be an exception to this, but all the "classical" dinosaurs, such as *T. rex* and *Triceratops*, would be missing.

11. Weiss 2016. Not everyone, I should add, accepts this inference. See Wade 2016.

12. One exception to the rule that one species can't give rise to a different species involves speciation via polyploidy, in which a mutation causes an organism to have more than two sets of chromosomes. We will consider another exception in our discussion of endosymbiosis in chapter 11.

13. This is what an emission spectrum looks like.

14. My trees were inspired by those used by biologist Kevin de Queiroz. See de Queiroz 2007, 882.

15. This appears to be the consensus view, but it is by no means universally held.

16. To *raft*, in this sense of the word, is to be involuntarily transported to an island on floating vegetation that is either blown by the wind or carried by currents.

17. "Romance Languages" 2017.

18. As we shall see in chapter 17, there are exceptions.

19. Since Nancy is Rh-negative, her Rh factor genes must be – –. (Since the Rh-factor gene is dominant, if you have even one + gene, you will be Rh-positive.) Since Paul is Rh-positive, his Rh-factor genes are either + + or + –. In the first case, any baby he and Nancy have will carry a + gene, meaning that it will be Rh-positive. In the second case, there is a one-in-two chance that their baby's Rh-factor genes will be – –, meaning that the baby will be Rh-negative.

20. Fortunately for women with Rh-negative blood, injections of Rho(D) immune globulin can prevent their immune system from being triggered by the presence in them of an Rh-positive fetus.

21. If the horse is male and the donkey is female, the offspring will be a hinny, which will also be infertile.

22. Lofholm 2007.

23. Gibbons 2011.

24. To better understand what this means, realize that if a *Homo sapiens* male mated with a Neanderthal female, their offspring would have 50 percent Neanderthal genes. If the offspring subsequently mated with a *Homo sapiens*, the percentage of Neanderthal genes would drop to 25 percent.

25. Reich 2010.

26. Another way for *Homo sapiens* to acquire Denisovan DNA is to mate with Neanderthals who carried it as a result of their ancestors having mated with Denisovans.

27. Those of Melanesian ancestry, who mostly inhabit New Guinea and the other islands northwest of Australia, carry unusually high amounts of Denisovan DNA in their genome. They also carry Neanderthal DNA. This is presumably a consequence of the route they took during their migration out of Africa and the adventures they had while migrating. See Vernot 2016.

28. Singer 2016a.

29. Huerta-Sánchez 2014.

30. Ackermann 2016.

CHAPTER 7

1. For a quick introduction to enzymes and to the catalytic role played by proteins, see Hobbs 2015.

2. Simple proteins can fold themselves. Complex proteins need the assistance of proteins called *chaperonins* to fold correctly.

3. Lane 2006, 10.

4. The recipe in a protein-coding gene can be edited to make many different proteins. An extreme example of this phenomenon can be found in fruit flies: one of their protein coding genes provides the recipe for more than 38,000 different proteins! And these "sibling" proteins, as they are called, can do different—even opposite—things. For a helpful discussion of this phenomenon, see Greenwood 2016.

5. Your body doesn't have a "protein-making machine." It instead has various biological components that work together to "read" DNA in order to construct molecules of protein. Some parts of this "machine" do their work inside the nucleus of the cell. Other parts function out in the cytoplasm, where the manufactured protein will be put to use.

6. Your body's protein-making machine actually employs not one but two codes, one to transcribe codons of DNA into strands of messenger RNA and a second to construct chains of amino acids on the basis of those strands. Figure 7.1 gives the "net effect" of applying these two codes.

7. This cartoon makes it look like proteins are manufactured by a single molecular "machine" that attaches to the DNA in the cell nucleus. What in fact happens, in eukaryotes like yourself, is that the enzyme RNA polymerase moves along the DNA strand and transcribes the information it finds there. The transcription takes the form of molecules of messenger RNA that then leave

the nucleus and head out into the cell, where ribosomes construct proteins in accordance with the "recipe" written on that messenger RNA.

8. There is one way to indicate methionine, there are four ways to indicate threonine, and there are six ways to indicate leucine. Notice that $1 \times 4 \times 4 \times 6 = 96$.

9. Bohannon 2016.

10. Zhang 2017.

11. If four nucleotides (A, T, C, and G) are available, a three-nucleotide-long codon can have any of $4 \times 4 \times 4 = 64$ different forms. With six nucleotides available, a codon can have $6 \times 6 \times 6 = 216$ different forms.

12. Loury 2012.

13. Wang, Haui, 2015.

14. Since codons are three nucleotides long, and since any of four nucleotides (A, T, C, and G) could appear in each of those three places, $4 \times 4 \times 4 = 64$ different codons are possible.

15. More precisely, there are 1.5×10^{84} different ways to pair up 64 codons with 20 amino acids and a "stop code." Yarus 2010, 163.

16. Notice that $2^{279} = 9.7 \times 10^{83}$, which is slightly less than 1.5×10^{84}.

17. It turns out that some genetic codes are better than others in their ability to prevent genetic mistakes from being made. See Zhu 2003. Since the evolutionary process "tries" to come up with efficient designs, we might therefore expect some similarities between two independently evolved genetic codes. Even taking this into account, though, there will almost certainly be extensive differences between any two such codes.

18. Elzanowski 2016.

CHAPTER 8

1. The identity of the first living organism is not known, of course. Furthermore, any attempt to identify it would depend on the definition of *life* that one favored. If we sent 10 biologists back in time, they might come up with 10 different candidates for the first living thing. But logic dictates that if, according to a particular definition of life, there were no living things at one time and there were living things at a later time, there must be a "first living thing" (or multiple simultaneous first living things) somewhere in between those times.

2. For a discussion of the possibility of life without RNA or DNA, see Benner 2004.

3. In chapter 7, we explored one way that genetic codes could differ—namely, in their assignments of amino acids to codons. They could also differ, though, in the nucleotides they use: as we have seen, adenine, thymine, cytosine, and guanine aren't the only possible nucleotides. Genetic codes can also differ in the length of their codons. Their "coding unit" might, for example, be four nucleotides long instead of three, or maybe only two nucleotides long. Furthermore, they could associate codons with different amino acids than the 20 that our code uses.

4. For a discussion of the shadow tree of life, see Davies 2007.

5. DNA, as we have seen, is used to store "recipes" that are written in nucleotides. When they are not being read, though, those nucleotides are paired with their complementary nucleotides—*A*s with *T*s, and *C*s with *G*s—to form the rungs of the twisted ladder that is the DNA double-helix. To say that your DNA is 3.2 billion base pairs long is to say that there are 3.2 billion of these rungs in your DNA recipe book. And one other comment is in order: your DNA recipe is stored not in a single chemical recipe book, but in the 46 different molecules known as chromosomes.

6. Zimmer, Carl, 2013.

7. Hutchison 2016.

8. Singer 2016b.

9. Aron 2015. Also see Extance 2016.

10. More precisely, ribosomes have a protein component.

11. Glasco 2016.

12. Although most genes are recipes for making proteins, tRNA genes are recipes for making the transfer RNA molecules that in essence carry the key to the genetic code, telling which codons "stand for" which amino acids.

13. This sort of RNA "self-replication" probably wouldn't involve a *single* RNA molecule making a copy of itself, but cooperative networks of such molecules making copies of themselves. For a discussion of this point, see Lehman 2015.

14. Cech 2012.

15. For a discussion of the role water plays in life processes, see Ball 2003.

16. For a useful discussion of these issues, see Martin 2014.

17. Science-minded readers may shudder on hearing me use the phrase "ancient astronauts." There are, to be sure, some colorful characters who advocate *directed* panspermia, but there are also serious scientists who have defended this theory. Francis Crick—who together with James Watson discovered the structure of DNA—was among them. See Crick 1973.

18. It took 33 years for the Voyager spacecraft, launched in 1977, to reach the edge of the solar system. If it continued to travel at its current speed and was headed in the right direction, it would take 93,000 years for it to reach Proxima Centauri.

CHAPTER 9

1. In vitro fertilization is an obvious exception.

2. Things are in fact a bit more complicated than this. The nuclei of the merged sperm and egg don't themselves "merge," in the usual sense of the word. Rather, they dissolve, so that the male and female chromosomes they contain can combine. A new nuclear envelope then forms around them. And a technical clarification: when the egg contains the nuclei of both the sperm and egg, these nuclei are referred to as *pronuclei*.

3. It is also possible, amazingly, for two identical twins to be of different sexes. This is because, both genetically and anatomically speaking, sex is far more complicated than most people think. For more information, see Tachon 2014.

4. Ainsworth 2015.

5. Ainsworth 2015.

6. In humans, there are exceptions to this. First and most obviously, an egg and sperm cell merge to make a zygote. It is also possible, though, for your muscle cells to merge to make one cell that has two nuclei—what is known as a *syncytium*.

7. Bianconi 2013. Actually, since your cells are constantly dying, your zygote had to divide into far more than 37 trillion cells in order for you to have the 37 trillion that you do. This last number, by the way, is open to debate. See Sender 2016.

8. There is a debate over whether women keep making eggs after they are born. For a non-technical discussion of this debate, see Dell'Amore 2012.

9. Spalding 2005.

CHAPTER 10

1. These cells would be the direct ancestors of the cells that comprise you and would therefore be the direct ancestors of "the cellular you."

2. For more on choanoflagellate colonies, see McGowan 2014.

3. Octopuses are one interesting exception to this. What you might naturally want to call an octopus's head is better described as its body. Its eyes and brain are between its tentacles and this "head." And this body design isn't the only strange thing about octopuses. They also have three hearts, filled with blood that, because it is copper based rather than iron based, is blue in color. The optic nerves and blood vessels of their eyes are positioned, quite sensibly, on the back of their retinas instead of the front, the way ours are. And one last thing: their esophagus runs through the middle of their brain, which in turn contains only about one third of their neurons, the rest being located in their tentacles.

4. Dayel 2011.

5. To gain a better understanding of this simple yet remarkable organism, see the WormAtlas.org website.

6. Scientists have discovered that by using a microscope-guided laser to zap a few of its neurons, they can cause a *C. elegans* nematode to lose interest in sex. Narayana 2016.

7. The exact number of cell types is not known, but ongoing initiatives to create a human cell atlas will give us much greater insight into what cell types there are and where the various types of cells reside within us. See Nowogrodzki 2017.

8. Rensberger 1996, 12-13.

9. It is an ability that they do not necessarily exercise. In most bee species, for example, bees maintain a solitary existence. See Singer 2014.

10. Ostwald 2016.

11. Your zygote was, more precisely, a *totipotent* stem cell. Like any *pluripotent* stem cell, it can give rise to any of the different specialized cells that now comprise you. What makes it special is that it can also give rise to an entire organism.

12. Coghlan 2014.

13. Slack 2014.

CHAPTER 11

1. For more on this discovery, see Irvine 2015.

2. There is considerable disagreement about when this event took place, with estimates varying between 2.1 and 1.5 billion years ago.

3. Ettema 2016, 39.

4. It may seem frivolous for me to be giving microbes proper names, but doing so helps us keep in mind that eukaryotes quite likely came to exist as the result of a singular event that involved two particular microbes.

5. Nick Lane (2006) makes the case for this, but not everyone agrees.

6. Archaea and bacteria aren't sexual organisms, so the pronoun *it* should, logically, be used in conjunction with them. I use *she* only to be consistent with the common practice of referring to the results of cell division as *daughter cells*. And yes, *Archie* is usually a male name—but not always.

7. Your blood cells are an exception.

8. Your mitochondria, by the way, are surrounded by not one but two membranes. This is sometimes cited as evidence of their endosymbiotic origin. Think about what happens when you push a finger into a balloon. The rubber of that balloon will wrap around the finger. The idea is that this sort of thing happened when Archie engulfed Becky. (In the second "frame" of Figure 11.1, we can see Archie's membrane surrounding Becky.) Some think that the inner membrane of your mitochondria can be traced back to Becky's outer membrane and that the outer membrane of your mitochondria can be traced back to Archie's outer membrane. Others, however, think that Becky herself, as an Alphaproteobacteria, would have had a double membrane, meaning that the inner and outer membranes of your mitochondria can be traced back to her. For a discussion of this debate, see Moran 2010.

9. Wang, Xu, 2015.

10. There are, as we have seen, exceptions to this claim. If you have any chimeric cells in your body, they won't be able to trace their ancestry back to your zygote, and if the Earth has a shadow tree of life—see Figure 8.1—not every currently existing cell will be able to trace its ancestry back to LUCA.

11. This view has been challenged. See Ankel-Simons 1996.

12. Zimmer, Marc, 2015, 29.

13. Alberts 2002, 770.

14. Lane 2006, 3, 11.

15. Many eukaryotes have much simpler genomes. The genome of the yeast in your kitchen, for example, has only 12 million base pairs.

16. Biologists have recently discovered that the mitochondria that are the source of much of your body's heat are themselves significantly warmer than your body. It is as if they create their own thermal microenvironment. See Le Page 2017.

17. Ettema 2016.

18. Fossil evidence indicates that the date could have been no later than 1.2 billion years ago. See Björn 2009.

19. Cyanobacteria are sometimes referred to as *algae*, but only eukaryotes can be algae in the strictest sense of the word.

20. Jukes 1990.

21. Cyanobacteria would still be around to produce it.

22. The oldest fossil evidence of sexual reproduction dates back 1.2 billion years. See Butterfield 2000.

CHAPTER 12

1. Chawla 2014.

2. Dickson 2015.

3. It has lately been discovered that archaea play a much bigger role in our gut biome than was formerly thought. See Raymann 2017.

4. Rose 2015.

5. The one-in-ten ratio has come into question. According to Sender 2016, the ratio is more like one-in-two, but in his cell census, he counts blood cells as bodily cells. This has a dramatic impact on the ratio, inasmuch as five in six of your cells are blood cells. Others would argue that

blood cells aren't really a part of you and therefore should be omitted from the cell census. This will dramatically reduce the ratio of bodily cells to "boarder" cells.

6. Manriquea 2016.

7. Grens 2014.

8. Gill 2011.

9. Nunes-Alves 2016.

10. Rogier 2014. See also Yong 2016.

11. David 2013. For a rather entertaining description of how gut biomes change over time, see Zeldovich 2014.

12. Alcock 2014.

13. Goldman 2016.

14. Goldman 2016.

15. Engelhaupt 2015.

16. Ehrenberg 2015.

17. Mole 2014.

18. Hamzelou 2015.

19. Ridaura 2013.

20. Hentschel 2012.

21. Kembel 2014.

22. "Plants Prepackage" 2014.

23. Brucker 2013.

24. Arnold 2014.

25. For a more detailed examination of life in a gnotobiotic world, see Gilbert 2014.

CHAPTER 13

1. Freitas 1998.

2. It is possible for mass to be converted directly into energy and for energy to be converted into mass, but in the ordinary course of things, such conversions are negligible.

3. One exception would be substances injected into you.

4. Normally, your body "burns" protein only if you are undernourished.

5. Most of the oxygen in our atmosphere is *molecular oxygen*, which is comprised of two oxygen atoms. It is also possible for oxygen to exist as a single atom, known as *atomic oxygen*, but it usually isn't long before such atoms combine to make molecular oxygen. And finally, it is possible for three atoms of oxygen to combine to form a molecule of *ozone*.

6. Muller 2012.

7. United States Department of Agriculture 2003, 14–21.

8. Zmuda 2011.

9. In thinking about these numbers, you should keep in mind that the water you drink is less than 1 percent of the 100 gallons of water you probably use each day, for bathing or showering, for flushing toilets, for watering the lawn, and so on.

10. Creager 2013, 31.

11. Aebersold 1954, 231–232.

CHAPTER 14

1. If we did a census of your atoms, 12 percent of them would be carbon. Because they are relatively heavy, though, these atoms make up 18.5 percent of your mass.

2. There is evidence, by the way, that plants have taken in a significant part of the carbon that we humans have placed into the atmosphere. See Zimmer, Carl, 2017.

3. Bellows 2008.

4. Biello 2008.

5. As a result of incomplete burning, there are also lesser amounts of other molecules, including CO, carbon monoxide.

6. Readers interested in the details of this process are invited to explore the Calvin cycle.

7. Reece 2014, 188.

8. Dillon 2012, 37.

9. The world recently produced about 1 trillion metric tons of corn. See United States Department of Agriculture 2017. To put this number into context, realize that very little of the corn that is grown is "sweet corn" that might appear on the dinner table. It is instead "field corn," most of which will be used to make either ethanol, for use as a gasoline additive, or to make animal feed.

10. Scientists discover what elements are essential for life by systematically depriving test animals of elements to see what happens. If removing an element from their diet results in their death, it is essential. This is the process by which they recently added bromine to the list of essential elements. See McCall 2014.

CHAPTER 15

1. This isn't the only question we can ask regarding the Big Bang. Indeed, in the last chapter of this book, we will ask and attempt to answer two more. First, why was there a Big Bang? And second, given that there *was* a Big Bang, why did it result in *this* universe, rather than *some other* universe, with different physical laws?

2. Realize that atomic nuclei don't want to merge. Because they are positively charged, they repel each other, and the repulsive force gets stronger the closer they get. It took the extremely high pressures that existed in the early universe to overcome this force.

3. How, one wonders, can there be *re*combination of electrons and nuclei if they hadn't yet combined? It is misleading in the way that the phrase "refried beans," in Mexican cooking, is.

4. Astrophysicists refer to this obstacle as the *mass-5* and *mass-8 bottleneck* in nucleosynthesis. For a somewhat technical discussion of this phenomenon, see "Nucleosynthesis" 1998–2018.

5. It is possible that you have hydrogen atoms that have never been in a star and therefore have not been blown through space in a supernova event.

6. "Earth's Gold" 2013. See also Sokol 2017.

7. Astronomers somewhat confusingly refer to this first population of stars as *Population III stars*.

8. "Astronomers Find Sun's 'Long-Lost Brother'" 2014.

CHAPTER 16

1. This is their composition by mass. Switch to composition by number of atoms, and 92 percent of them are hydrogen.

2. Love 2004.

3. Valley 2014.

4. But what if some lead got incorporated into the crystal along with the uranium? Wouldn't this make the age calculation inaccurate? Scientists have determined that because of chemical differences between lead and uranium, lead cannot substitute for zircon in crystal formation, meaning that any lead found in a zircon must be the result of uranium decay.

5. Rumble 2013.

6. Genge 2017.

7. Willbold 2011.

8. Some of the earth's mountains are the result not of plate tectonics but mantle plumes. These plumes are the result of upwelling of hot rock within the earth's mantle. The volcanoes of Hawaii, which are in the middle of a tectonic plate, are thought to be the result of one such plume.

9. The earth is estimated to have 1.386×10^9 km^3 of water and a surface area of 5.1×10^8 km^2. The average depth of the ocean on a marble-smooth earth would therefore be $(1.386 \times 10^9 \text{ km}^3)/(5.1 \times 10^8 \text{ km}^2) = 2.7$ km, or about 9,000 feet.

10. "Abundance in Earth's Crust of the Elements" n.d.

11. Mikhail 2014.

CHAPTER 17

1. This is a simplification, since besides the three common variants I describe, there are dozens of uncommon variants.

2. Chi 2016.

3. In eukaryotes, genes are typically divided into regions known as *introns* and *exons*. The introns can be thought of as junk DNA within a gene. They will be edited out by the complex molecular machines known as *spliceosomes* before protein construction begins. We have seen that less than 2 percent of your DNA codes for proteins. Only a small portion of this 2 percent, however, will be located in the exonic regions of your protein-coding genes. The rest will be edited out. Conclusion: if we think of your genome as a cookbook for making proteins, it will be a very strange cookbook. Only 2 percent of its pages will contain recipes, and those recipes will have to be severely edited before they will be usable. Although the presence within us of introns might seem strange, it has its genetic advantages. The division of a gene into multiple exons gives your spliceosomes the option of including some exons and omitting others. This is why one gene can give rise to multiple proteins.

4. Nee 2016.

5. Like any generalization boldly made about biology, this one has exceptions: the tips of a male's X and Y chromosomes can recombine. See Hinch 2014.

6. Notice that 1 egg per month over a 40-year reproductive life is 480 eggs. Notice that 100,000,000 sperm per day / 86,400 seconds per day = 1,157 sperm per second. And finally, notice that if a man produces an average of 100,000,000 sperm per day over a 60-year reproductive life, he will produce a total of $100{,}000{,}000 \times 365 \times 60 = 2{,}190{,}000{,}000{,}000$ sperm.

7. Actually, minor alterations are possible. As mentioned in the previous endnote, crossover can take place between the sex chromosomes that your father inherited from his parents, but it affects only the tips of those chromosomes, meaning that most Y-chromosome DNA is passed on unchanged from father to son. It is also possible for mutations to change the middle part of the Y chromosome that the father passes on to his son.

8. Prüfer 2012.

9. Fan 2002.

10. By looking at the genes shared by living things, we can gain further insight into LUCA, the assumption being that the most likely way genes would be widespread is if they had been passed down, with minimal change, from LUCA. Recently, scientists identified 355 of these "highly conserved genes" and on the basis of what they found, concluded that LUCA probably lived in a thermal vent at the bottom of the ocean. See Weiss 2016.

11. Magadum 2013.

12. Meyer 2017.

13. Harris 1991.

14. Arnold 2016.

15. Dawkins 1976.

16. Worker bees sometimes lay unfertilized eggs that develop into the male drones that mate with fertile queen bees.

CHAPTER 18

1. Bertone 2016.

2. Calling them *centipedes* suggests that they have a hundred feet, or maybe a hundred pairs of legs. A typical house centipede will have fifteen pairs of legs, but some species have hundreds. And while millipedes routinely have more legs than centipedes, no millipede has a thousand legs.

3. Microbiologist Laura A. Hug describes meadows as "one of the most microbially complex environments on the planet." Zimmer, Carl, 2016.

4. The compound 2-methylisoborneol also contributes to the after-rain smell. Ozone molecules, by the way, are responsible for the imminent-rain smell that might precede a thunderstorm.

5. See Meganathan n.d. and "The chemical compounds behind the smell of rain" n.d.

6. Clarke 2002.

7. Borgonie 2011.

8. Barras 2013.

9. DeLeon-Rodriguez 2013.

10. Fox 2015.

11. Engelhaupt 2016.

12. Grossman 2010.

13. Trager 2016.

14. Gombay 2016.

15. For a discussion of this "Great Plate Count Anomaly," as it is called, see Lewis 2010.

16. Ratnarajah 2014. See also Monbiot 2014.

17. Dodd 2008.

18. Kirkby 2016.

19. A geologist will argue that because coal's carbon is derived from organic sources, it does not count as a mineral. A lawyer, however, will likely disagree. See, for example, Reservation of Coal and Mineral Rights, 43 U.S.C. §299 (1993).

20. Hazen 2014.

CHAPTER 19

1. In John 11:25, Jesus is quoted as saying that "I am the resurrection and the life. Whoever believes in me, though he die, yet shall he live." Christians tend to speak in terms not of their mind but of their soul or spirit. It is unclear, however, what difference, if any, there is between a soul and a spirit. It is also unclear what either of these entities is, other than a mind.

2. 1 Corinthians 15: 42–53.

3. Those who believe in reincarnation might claim that it is your soul or consciousness that comes back. It isn't clear what these are, if not what you would normally think of as your mind.

4. Buswell 2013: 49–50, 708–709.

5. I explore this phenomenon in detail in my *On Desire: Why We Want What We Want* (2006).

6. Graziano 2016.

7. It isn't clear how he got permission to do these disinterments—or for that matter, whether he even asked for permission. Perhaps the cemetery was being moved?

8. Motter 1898, 203.

9. Palmer 2014.

10. Loeffler 2002, 162.

11. Hamilton 1996, 87.

12. Coyne 2012.

13. A mother's and father's contributions to the creation of a zygote are not equal. All the father contributes is 23 strands of DNA. The mother contributes that together with the rest of a functional cell. She also contributes, as we have seen, the DNA of that cell's mitochondria. The primary role played by the egg cell becomes clearer when we realize that under exceedingly rare circumstances, it can develop into a person without the addition of DNA from a man. See Strain 1995. A case can therefore be made that it is really the mother whose cells live on after her death and that the father simply plays a role in facilitating that cellular immortality.

14. As we have seen, you carry two different sets of DNA, one in the nucleus of your cells and the other in your mitochondria. A complete description of your genome will contain information about both sorts of DNA.

15. No person has been successfully cloned yet, but it is quite likely that someday, someone will be. And regarding cloning, a comment is in order: to get a "perfect" clone of yourself, you will need to copy both your nuclear and your mitochondrial DNA.

CHAPTER 20

1. In 1983, the definition of a meter was tweaked so that this number would come out exact. More precisely, a meter was defined as the distance light travels in vacuum in 1/299,792,458 of a second.

2. Smolin 1979, 39.

3. Rig Veda 10:121.

4. In the eleventh century, the monk Gaunilo of Marmoutiers attacked the ontological argument by pointing out that if, following St. Anselm, we characterize "the Perfect Island" as *the island than which no greater island can be imagined*, we will be able to prove that the Perfect Island exists—which it obviously doesn't. Logicians have subsequently suggested that the ontological

argument goes astray by assuming that existence is a property, the way omniscience and omnipotence are, when it is instead a condition that must obtain before properties can be had.

5. In the second chapter of Genesis, we are told that God put Adam in the Garden of Eden (Genesis 2:8), implying that the garden existed before Adam did. We are also told that God's reason for putting Adam there was so that he could work the garden and take care of it. See Genesis 2: 15.

6. Exodus 16:3.

7. Bostrom 2003.

8. The sky is blue because the atmosphere's N_2 and O_2 molecules, due to their size, scatter blue light. When you look where the sun isn't, you see this scattered light, so the sky looks blue. But of course, this answer doesn't get to the bottom of things. We are left, for example, with the question of why the atmosphere has N_2 and O_2 molecules, why light scatters, and why molecules and light exist, to begin with.

WORKS CITED

"Abundance in Earthwebsite's Crust of the Elements." N.d. Wolfram Research's PeriodicTable website, http://periodictable.com/Properties/A/CrustAbundance.v.log.html.

Ackermann, Rebecca Rogers, et al. 2016. "The Hybrid Origin of 'Modern' Humans." *Evolutionary Biology* 41 (1): 1–11.

Aebersold, Paul C. 1954. "Radioisotopes—New Keys to Knowledge." *Annual Report of the Board of Regents of the Smithsonian Institution: 1953*. Washington, DC: United States Government Printing Office, 219–240.

Ainsworth, Claire. 2015. "Sex Redefined." *Nature* 518 (7539): 288–291.

Alberts, Bruce, et al. 2002. *Molecular Biology of the Cell*. 4th edition. New York: Garland Science.

Alcock, Joe, et al. 2014. "Is Eating Behavior Manipulated by the Gastrointestinal Microbiota? Evolutionary Pressures and Potential Mechanisms." *BioEssays* 36 (10): 940–949.

Alexander, Robert M. 2004. "Bipedal Animals, and Their Differences from Humans." *Journal of Anatomy* 204 (5): 321–330.

Ankel-Simons, Friderun, et al. 1996. "Misconceptions about Mitochondria and Mammalian Fertilization: Implications for Theories on Human Evolution." *PNAS* 93 (24): 13859–13863.

Arnold, Carrie. 2014. "Evolving with a Little Help from Our Friends." *Quanta Magazine* website, June 4, https://www.quantamagazine.org/20140604-evolving-with-a-little-help-from-our-friends/.

Arnold, Carrie. 2016. "Virus Pumps up Male Muscles—In Mice." *Nature News* website, September 12, http://www.nature.com/news/virus-pumps-up-male-muscles-in-mice-1.20574.

Aron, Jacob. 2015. "DNA in Glass—The Ultimate Archive." *New Scientist* 225 (3008): 15.

"Astronomers Find Sun's 'Long-Lost Brother,' Pave Way for Family Reunion." 2014. *Astronomy Magazine* website, May 13, http://www.astronomy.com/news/2014/05/astronomers-find-suns-long-lost- brother-pave-way-for-family-reunion.

Balaresque, Patricia, et al. 2015. "Y-chromosome Descent Clusters and Male Differential Reproductive Success: Young Lineage Expansions Dominate Asian Pastoral Nomadic Populations." *European Journal of Human Genetics* 23: 1413–1422.

Ball, Philip. 2003. "Water: The Molecule of Life." An Interview by "Astrobio" for *Astrobiology Magazine,* May 7, https://www.astrobio.net/origin-and-evolution-of-life/water-the-molecule-of-life/.

Balter, Michael. 2014. "How Farming Reshaped Our Genomes." *Science* website, January 26, http://www.sciencemag.org/news/2014/01/how-farming-reshaped-our-genomes.

Barras, Colin. 2013. "Deep Life: Biology's Final Frontier." *New Scientist* 218 (2914): 36–39.

Barton, Fiona. 2008. "Shock for the Married Couple Who Discovered They Are Twins Separated at Birth." *Daily Mail* website, January 11, http://www.dailymail.co.uk/news/article-507588/Shock-married-couple-discovered-twins-separated-birth.html.

Bellows, Sierra. 2008. "The Hair Detective." *University of Virginia Magazine*, Summer, http://uvamagazine.org/articles/the_hair_detective/.

Benner, Steven A., et al. 2004. "Is There a Common Chemical Model for Life in the Universe?" *Current Opinion in Chemical Biology* 8 (6): 672–689.

Bertone, Matthew A., et al. 2016. "Arthropods of the Great Indoors: Characterizing Diversity inside Urban and Suburban Homes." *PeerJ,* January 19, https://peerj.com/articles/1582/.

Bianconi, Eva, et al. 2013. "An Estimation of the Number of Cells in the Human Body." *Annals of Human Biology* 40 (6): 463–471.

Biello, David. 2008. "That Burger You're Eating Is Mostly Corn." *Scientific American* website, November 12, http://www.scientificamerican.com/article/that-burger-youre-eating-is-mostly-corn/.

Björn, Lars Olof, and G. Govindjee. 2009. "The Evolution of Photosynthesis and Chloroplasts." *Current Science* 96 (11): 1466–1474.

Bohannon, John. 2016. "Biologists Are Close to Reinventing the Genetic Code of Life." *Science* website, August 18, http://www.sciencemag.org/news/2016/08/biologists-are-close-reinventing-genetic-code-life.

Borgonie, Gaetan, et al. 2011. "Nematoda from the Terrestrial Deep Subsurface of South Africa." *Nature* 474 (7349): 79–82.

Bostrom, Nick. 2003. "Are You Living in a Computer Simulation?" *Philosophical Quarterly* 53 (211): 243–255.

Brucker, Robert, and Seth Bordenstein. 2013. "The Hologenomic Basis of Speciation: Gut Bacteria Cause Hybrid Lethality in the Genus *Nasonia*." *Science* 341 (6146): 667–669.

Buswell, Robert E., and Donald S. Lopez. 2013. *The Princeton Dictionary of Buddhism*. Princeton, NJ: Princeton University Press.

Butterfield, Nicholas J. 2000. "*Bangiomorpha pubescens* n. gen., n. sp.: Implications for the Evolution of Sex, Multicellularity, and the Mesoproterozoic/Neoproterozoic Radiation of Eukaryotes." *Paleobiology* 26 (3): 386–404.

Carter, Chelsea J. 2012. "Secret Revealed: Ohio Woman Unknowingly Married Father." CNN website, September 23, http://www.cnn.com/2012/09/21/us/ohio-woman-marries-father/.

Cassidy, Lara M., et al. 2016. "Neolithic and Bronze Age Migration to Ireland and Establishment of the Insular Atlantic Genome." *PNAS* 113 (2): 368–373

Chatterjee, Prata. 2015. "Drone Pilots Are Quitting in Record Numbers." *Mother Jones* website, March 5, http://www.motherjones.com/politics/2015/03/drone-pilots-are-quitting-record-numbers.

Chawla, Dalmeet Singh. 2014. "Bacteria on Pubic Hair Could Be Used to Identify Rapists." *Science* website, December 15, http://www.sciencemag.org/news/2014/12/bacteria-pubic-hair-could-be-used-identify-rapists.

Cech, Thomas R. 2012. "The RNA Worlds in Context." *Cold Spring Harbor Perspectives in Biology* 4 (7): a006742.

"The Chemical Compounds behind the Smell of Rain." N.d. Compoundchem website, http://www.compoundchem.com/2014/05/14/thesmellofrain/.

Chi, Kelly Rae. 2016. "The Dark Side of the Human Genome." *Nature* 538 (7624): 275–277.

Christoforou, Peter 2014. "5 Bizarre Paradoxes Of Time Travel Explained." Astronomy Trek website, December 20, http://www.astronomytrek.com/5-bizarre-paradoxes-of-time-travel-explained/.

Clarke, Tom. 2002. "Goldmine Yields Clues for Life on Mars." *Nature* website, December 9, http://www.nature.com/news/2002/021209/full/news021209-1.html.

Coghlan, Andy. 2014. "Amaze Balls: Testicles Site of Most Diverse Proteins." *New Scientist* website, November 6, https://www.newscientist.com/article/dn26506-amaze-balls-testicles-site-of-most-diverse-proteins/.

Conger, Krista. 2012. "Genetic Analysis of Ancient 'Iceman' Mummy Traces Ancestry from Alps to Mediterranean Isle." Stanford Medicine news center website, March 12, https://med.stanford.edu/news/all-news/2012/03/genetic-analysis-of-ancient-iceman-mummy-traces-ancestry-from-alps-to-mediterranean-isle.html.

Connor, Steve. 2008. "There's Nothing Wrong with Cousins Getting Married, Scientists Say." *Independent* website, December 23, http://www.independent.co.uk/news/science/theres-nothing-wrong-with-cousins-getting-married-scientists-say-1210072.html.

Coyne, Jerry A. 2012. "A Visit to the Grave of W. D. Hamilton." *Why Evolution Is True* blog, September 16, https://whyevolutionistrue.wordpress.com/2012/09/16/a-visit-to-the-grave-of-w-d-hamilton/.

Creager, Angela N. H. 2013. *Life Atomic: A History of Radioisotopes in Science and Medicine.* Chicago: University of Chicago Press.

Crick, Francis, and Leslie Orgel. 1973. "Directed Panspermia." *Icarus* 19: 341–346.

Curnoe, Darren. 2016. "When Humans Split from the Apes." *The Conversation* website, February 21, https://theconversation.com/when-humans-split-from-the-apes-55104.

David, Lawrence A., et al. 2013. "Diet Rapidly and Reproducibly Alters the Human Gut Microbiome." *Nature* 505 (7484): 559–563.

Davies, Paul. 2007. "Are Aliens among Us?" *Scientific American*, 297 (December): 62–69.

Dawkins, Richard. 1976. *The Selfish Gene.* Oxford, UK: Oxford University Press.

Dayel, Mark J., et al. 2011. "Cell Differentiation and Morphogenesis in the Colony-Forming Choanoflagellate *Salpingoeca rosetta*." *Developmental Biology* 357 (1): 73–82.

de Graciansky, Pierre-Charles, et al. 2011. *The Western Alps, from Rift to Passive Margin to Orogenic Belt*. New York: Elsevier.

DeLeon-Rodriguez, Natasha. 2013. "Microbiome of the Upper Troposphere: Species Composition and Prevalence, Effects of Tropical Storms, and Atmospheric Implications." *PNAS* 110 (7): 2575–2580.

Dell'Amore, Christine. 2012. "Women Can Make New Eggs After All, Stem-Cell Study Hints." *National Geographic* website, March 1, http://news.nationalgeographic.com/news/2012/02/120229-women-health-ovaries-eggs-reproduction-science/.

de Queiroz, Kevin. 2007. "Toward an Integrated System of Clade Names." *Systematic Biology*, 56 (6): 956–974.

Dickson, Robert P., and Gary B. Huffnagle. 2015. "The Lung Microbiome: New Principles for Respiratory Bacteriology in Health and Disease." *PLoS Pathogens*. 11 (7), July 9, http://journals.plos.org/plospathogens/article?id=10.1371/journal.ppat.1004923.

Dillon, Patrick F. 2012. *Biophysics: A Physiological Approach*. Cambridge, UK: Cambridge University Press.

Dodd, Scott. 2008. "DMS: The Climate Gas You've Never Heard Of." *Oceanus* 46 (3).

Durant, Will, and Ariel Durant. 1963. *The Story of Civilization: The Age of Louis XIV, 1648–1715*. New York: Simon and Schuster.

"Earth's Gold Came from Colliding Dead Stars." 2013. Harvard-Smithsonian Center for Astrophysics website, July 17, https://www.cfa.harvard.edu/news/2013-19.

Ehrenberg, Rachel. 2015. "Microbes May Be a Forensic Tool for Time of Death." *ScienceNews* blog, July 22, https://www.sciencenews.org/blog/culture-beaker/microbes-may-be-forensic-tool-time-death.

Elhaik, Eran. 2014. "Geographic Population Structure Analysis of Worldwide Human Populations Infers Their Biogeographical Origins." *Nature Communications* 5: 3513.

Elzanowski, Andrzej (Anjay), and Jim Ostell. 2016. "The Genetic Codes." National Center for Biological Information website, April 30, http://www.ncbi.nlm.nih.gov/Taxonomy/Utils/wprintgc.cgi.

Engelhaupt, Erika. 2015. "You're Surrounded by Bacteria That Are Waiting for You to Die." *Gory Details* blog, December 12, http://phenomena.nationalgeographic.com/2015/12/12/youre-surrounded-by-bacteria-that-are-waiting-for-you-to-die/.

Engelhaupt, Erika. 2016. "See the Ugly Beauty That Lives in a Toxic Cave. *Gory Details* blog, June 6, http://phenomena.nationalgeographic.com/2016/06/03/see-the-ugly-beauty-that-lives-in-a-toxic-cave/.

Ettema, Thijs J. G. 2016. "Mitochondria in the Second Act." *Nature* 531 (7592): 39–40.

Extance, Andy. 2016. "How DNA Could Store All the World's Data." *Nature* 537 (7618): 22–24.

Fan, Yuxin. 2002. "Genomic Structure and Evolution of the Ancestral Chromosome Fusion Site in 2q13–2q14.1 and Paralogous Regions on Other Human Chromosomes." *Genome Research* 12 (11): 1651–1662.

Fields, R. Douglas. 2008. "Are Whales Smarter Than We Are?" *Scientific American* news blog, January 15, http://blogs.scientificamerican.com/news-blog/are-whales-smarter-than-we-are/.

Fox, Douglas. 2015. "Scientists Drill through 2,400 Feet of Antarctic Ice for Climate Clues." *Scientific American* website, January 16, https://www.scientificamerican.com/article/scientists-drill-through-2-400-feet-of-antarctic-ice-for-climate-clues/.

Frazer, Jennifer. 2015. "Two-Billion-Year-Old Fossils Reveal Strange and Puzzling Forms." *Scientific American* news blog, January 29, http://blogs.scientificamerican.com/artful-amoeba/two-billion-year-old-fossils-reveal-strange-and-puzzling-forms/.

Freitas, Robert A., Jr. 1998. "Nanomedicine." Foresight Institute website, http://www.foresight.org/Nanomedicine/Ch03_1.html.

"Generation Length." 2015. International Society of Genetic Genealogy wiki, January 15, http://isogg.org/wiki/Generation_length.

Genge, Matthew J., et al. 2017. "An Urban Collection of Modern-Day Large Micrometeorites: Evidence for Variations in the Extraterrestrial Dust Flux through the Quaternary." *Geology* 45 (2): 119-122.

Gibbons, Ann. 2011. "The Species Problem." *Science* 331 (6013): 394.

Gibbons, Ann. 2015a. "How Europeans Evolved White Skin." *Science* website, April 2, http://www.sciencemag.org/news/2015/04/how-europeans-evolved-white-skin.

Gibbons, Ann. 2015b. "Trove of Teeth from Cave Represents Oldest Modern Humans in China." *Science* 350 (6258): 264.

Gilbert, Jack A., and Josh D. Neufeld. 2014. "Life in a World without Microbes." *PLoS Biology* 12 (12), December 16, http://journals.plos.org/plosbiology/article?id=10.1371%2Fjournal.pbio.1002020.

Gill, Erin E., and Fiona S. L. Brinkman. 2011. "The Proportional Lack of Archaeal Pathogens: Do Viruses/Phages Hold the Key?" *BioEssays* 33 (4): 248–254.

Glasco, Derrick M. 2016. "Beyond the DNA-Protein Paradox: A 'Clutch' of Other Chicken-Egg Paradoxes in Cell and Molecular Biology." *Answers Research Journal* 9: 209–227.

Goldman, Bruce. 2016. "Gut Bust: Intestinal Microbes in Peril." Stanford Medicine website, http://stanmed.stanford.edu/2016spring/gut-bust.html.

Gombay, Katherine. 2016. "Nearing the Limits of Life on Earth." McGill University website, January 19, https://www.mcgill.ca/newsroom/channels/news/nearing-limits-life-earth-257865.

Gordon, Kara. 2014. "The Pope's Views on Evolution Haven't Really Evolved." *The Atlantic* website, October 30, http://www.theatlantic.com/national/archive/2014/10/pope-francis-evolution/382143/.

Graziano, Michael. 2016. "Why You Should Believe in the Digital Afterlife." *The Atlantic* website, July 14, https://www.theatlantic.com/science/archive/2016/07/what-a-digital-afterlife-would-be-like/491105/.

Green, Richard E., et al. 2010. "A Draft Sequence of the Neandertal Genome." *Science* 328 (5979): 710–722.

Greenwood, Veronique. 2016. "A Secret Flexibility Found in Life's Blueprints." *Quanta Magazine* website, April 26, https://www.quantamagazine.org/20160426-one-gene-many-proteins/.

Grens, Kerry. 2014. "The Maternal Microbiome." *The Scientist* website, May 21, http://www.the-scientist.com/?articles.view/articleNo/40038/title/The-Maternal-Microbiome/.

Grossman, Lisa. 2010. "Underground Oasis Found Below Earth's Driest Desert." *New Scientist* website, February 18, https://www.newscientist.com/article/dn21497-underground-oasis-found-below-earths-driest-desert/.

Hales, Dinah F. 2002. "Lack of Detectable Genetic Recombination on the X Chromosome During the Parthenogenetic Production of Female and Male Aphids." *Genetics Research* 79: 203–209.

Hamilton, W. D. 1996. "My Intended Burial and Why." In *Narrow Roads of Gene Land: The Collected Papers of W. D. Hamilton*. Volume 3. New York: W. H. Freeman.

Hamzelou, Jessica. 2015. "Don't Give Me That Crap." *New Scientist* 225 (3008): 8–9.

Harris, J.R. 1991. "Hypothesis: The Evolution of Placental Mammals." *FEBS Letters* 295 (1–3): 3–4.

Haub, Carl. 2011. "How Many People Have Ever Lived on Earth?" Population Research Bureau website, http://www.prb.org/Publications/Articles/2002/HowManyPeopleHaveEverLivedonEarth.aspx.

Hazen, Robert. 2014. "Mineral Fodder." *Aeon* website, June 24, https://aeon.co/essays/how-life-made-the-earth-into-a-cosmic-marvel.

Hein, Jotun. 2004. "Human Evolution: Pedigrees for All Humanity." *Nature* 431 (7008): 518–519.

Hentschel, Ute, et al. 2012. "Genomic Insights into the Marine Sponge Microbiome." *Nature Reviews Microbiology* 10 (9): 641–654.

Hinch, Anjali G., et al. 2014. "Recombination in the Human Pseudoautosomal Region PAR1." *PLoS Genetics* 10 (7), July 17, http://journals.plos.org/plosgenetics/article?id=10.1371/journal.pgen.1004503.

Hobbs, Bernie. 2015. "Chemistry: Not As Easy as A + B → C." ABC Science website, May 25, http://www.abc.net.au/science/articles/2015/05/25/4229949.htm.

Horan, Richard D., et al. 2005. "How Trade Saved Humanity from Biological Exclusion: An Economic Theory of Neanderthal Extinction." *Journal of Economic Behavior & Organization* 58 (1): 1–29.

Hublin, Jean-Jacques, et al. 2017. "New Fossils from Jebel Irhoud, Morocco and the Pan-African Origin of *Homo sapiens*." *Nature* 546 (7657): 289–292.

Huerta-Sánchez, Emilia, et al. 2014. "Altitude Adaptation in Tibetans Caused by Introgression of Denisovan-like DNA." *Nature* 512 (7513): 194–197.

Hutchison, Clyde A., III, et al. 2016. "Design and Synthesis of a Minimal Bacterial Genome." *Science* 351 (6280): 1414–1424.

Irvine, William B. 2006. *On Desire: Why We Want What We Want.* New York: Oxford University Press.

Irvine, William B. 2015. *Aha! The Moments of Insight That Shape Our World.* New York: Oxford University Press.

Jacob, Francois. 1977. "Evolution and Tinkering." *Science* 196 (4295): 1161–1166.

Jukes, T. H., and S. Osawa. 1990. "The Genetic Code in Mitochondria and Chloroplasts." *Experientia* 46 (11–12): 1117–1126.

Kembel, Steven W., et al. 2014. "Relationships between Phyllosphere Bacterial Communities and Plant Functional Traits in a Neotropical Forest." *PNAS* 111 (38): 13715–13720.

Kindlmann, Pavel, et al. 1989. "Developmental Constraints in the Evolution of Reproductive Strategies: Telescoping of Generations in Parthenogenetic Aphids." *Functional Ecology* 3 (5): 531–537.

Kirkby, Jasper, et al. 2016. "Ion-induced Nucleation of Pure Biogenic Particles." *Nature* 533 (7604): 521–526.

Kirschner, Marc W., et al. 2005. *The Plausibility of Life: Resolving Darwin's Dilemma.* New Haven, CT: Yale University Press.

Kolbe, J. J., et al. 2012. "Founder Effects Persist Despite Adaptive Differentiation: A Field Experiment with Lizards." *Science* 335: 1086–1089.

Kuban, Glen J. N.d. "The Texas Dinosaur/'Man Track' Controversy." The TalkOrigins website, http://www.talkorigins.org/faqs/paluxy.html.

Lane, Nick. 2006. *Power, Sex, Suicide: Mitochondria and the Meaning of Life.* Oxford, UK: Oxford University Press.

Lane, Nick. 2009. *Life Ascending: The Ten Great Inventions of Evolution.* New York: W. W. Norton.

Lanier, Shannon. 2000. "Book Discussion on Jefferson's Children: The Story of One American Family." An interview with C-SPAN, October 16, http://www.c-span.org/video/?160092-1/book-discussion-jeffersons-children-story-one-american-family.

Lehman, Niles. 2015. "RNA Self-assembly: Cooperation at the Origins of Life." YouTube video, March 25, https://www.youtube.com/watch?v=vrpADqF3VBo.

Le Page, Michael. 2017. "The Energy Generators inside Our Cells Reach a Sizzling 50°C." *New Scientist* website, May 4, https://www.newscientist.com/article/2129849-the-energy-generators-inside-our-cells-reach-a-sizzling-50c/.

Lewis, Kim. 2010. "The Uncultured Bacteria." *Small Things Considered* blog, July 12, http://schaechter.asmblog.org/schaechter/2010/07/the-uncultured-bacteria.html.

Llorente, M. Gallego, et al. 2015. "Ancient Ethiopian Genome Reveals Extensive Eurasian Admixture in Eastern Africa." *Science* 350 (6262): 820–822.

Loeffler, Jack. 2002. *Adventures with Ed: A Portrait of Abbey*. Albuquerque: University of New Mexico Press.

Lofholm, Nancy. 2007. "Mule's Foal Fools Genetics with 'Impossible' Birth." *Denver Post* website, July 26, http://www.denverpost.com/news/ci_6464853/mule-foal-fools-genetics-impossible-birth.

Loury, Erin. 2012. "The Origin of Blond Afros in Melanesia." *Science* website, May 3, http://www.sciencemag.org/news/2012/05/origin-blond-afros-melanesia.

Love, Stanley G. and Donald R. Pettit. 2004. "Fast, Repeatable Clumping of Solid Particles in Microgravity." *Lunar and Planetary Science* 35: 1119.

Lyons, Timothy W., et al. 2014. "The Rise of Oxygen in Earth's Early Ocean and Atmosphere." *Nature* 506 (7488): 307–315.

Magadum, Santoshkumar, et al. 2013. "Gene Duplication as a Major Force in Evolution." *Journal of Genetics* 92 (1): 155–161.

Main, Douglas. 2014. "Galapagos Giant Tortoise Brought Back from Brink of Extinction." *Newsweek* website, October 28, http://www.newsweek.com/galapagos-giant-tortoise-brought-back-brink-extinction-280593.

Manriquea, Pilar, et al. 2016. "Healthy Human Gut Phageome." *PNAS* 113 (37): 10400–10405

Marino, Lori, et al. 2007. "Cetaceans Have Complex Brains for Complex Cognition." *PLoS Biology* 5 (5), May 15, http://journals.plos.org/plosbiology/article?id=10.1371/journal.pbio.0050139.

Martin, William F., et al. 2014. "Energy at Life's Origin." *Science* 344 (6188): 1092–1093.

McCall, A. Scott, et al. 2014. "Bromine Is an Essential Trace Element for Assembly of Collagen IV Scaffolds in Tissue Development and Architecture." *Cell* 157 (6): 1380–1392.

McGowan, Kat. 2014. "Where Animals Come From." *Quanta Magazine* website, July 29, https://www.quantamagazine.org/20140729-where-animals-come-from/.

Meganathan, Rangaswamy. N.d. "What Causes the Characteristic Smell of Soil?" Northern Illinois University website, http://niu.edu/biology/about/faculty/meganathan/smell-of-soil.shtml.

Meyer, Thomas J., et al. 2017. "Endogenous Retroviruses: With Us and against Us." *Frontiers in Chemistry* website, April 7, https://www.ncbi.nlm.nih.gov/pubmed/28439515.

Mikhail, Sami, and Dimitri A. Sverjensky. 2014. "Nitrogen Speciation in Upper Mantle Fluids and the Origin of Earth's Nitrogen-Rich Atmosphere." *Nature Geoscience* 7: 816–819.

Mole, Beth. 2014. "Triclosan May Spoil Wastewater Treatment." *ScienceNews* website, June 19, https://www.sciencenews.org/article/triclosan-may-spoil-wastewater-treatment.

Monbiot, George. 2014. "Why Whale Poo Matters." *Guardian* website, December 12, https://www.theguardian.com/environment/georgemonbiot/2014/dec/12/how-whale-poo-is-connected-to-climate-and-our-lives.

Mora, Camilo, et al. 2011. "How Many Species Are There on Earth and in the Ocean?" *PLoS Biology*, August 23, http://journals.plos.org/plosbiology/article/asset?id=10.1371%2Fjournal.pbio.1001127.PDF.

Moran, Laurence A. 2010. "On the Origin of the Double Membrane in Mitochondria and Chloroplasts." *Sandwalk: Strolling with a Skeptical Biochemist* website, http://sandwalk.blogspot.com/2010/06/on-origin-of-double-membrane-in.html.

Morgan, M. H., et al. 2013. "Protective Buttressing of the Human Fist and the Evolution of Hominin Hands." *Journal of Experimental Biology* 216 (2): 236–244.

Motter, Murray G. 1898. "A Contribution to the Study of the Fauna of the Grave." *Journal of the New York Entomological Society*, 6 (4): 201–231.

Muller, Derek. 2012. "Are You Lightest in the Morning?" *Veratisium: An Element of Truth* blog, February 11, https://www.youtube.com/watch?v=lL2e0rWvjKI.

"Names." N.d. Open Domesday website, http://opendomesday.org/name/?indexChar=R.

Narayana, Anusha, et al. 2016. "Contrasting Responses within a Single Neuron Class Enable Sex-Specific Attraction in *Caenorhabditis elegans*." *PNAS* 113 (10): E1392–E1401.

Nee, Sean. 2016. "How Many Genes Does It Take to Make a Human?" Real Clear Science website, October 18, http://www.realclearscience.com/articles/2016/10/19/how_many_genes_does_it_take_to_make_a_human_109785.html.

"New App Urges Icelanders to 'Bump the App Before You Bump in Bed.'" 2013. Gadgets website, April 19, http://gadgets.ndtv.com/apps/news/new-app-urges-icelanders-to-bump-the-app-before-you-bump-in-bed-356344.

"Nomenclature of Inbred Mice." N.d. Jackson Laboratory website, https://www.jax.org/jax-mice-and-services/customer-support/technical-support/genetics-and-nomenclature/inbred-mice.

"Not Smith and Jones—Rare British Surnames on the Cusp of Extinction." 2011. *My Heritage* blog, April 26, http://blog.myheritage.com/2011/04/rare-british-surnames/.

Nowogrodzki, Anna. 2017. "How to Build a Human Cell Atlas." *Nature* 547 (7661): 24–26.

"Nucleosynthesis." 1998–2018. *The Physics Hypertextbook* website, http://physics.info/nucleosynthesis/.

Nunes-Alves, Cláudio. 2016. "Add the Microbiota to Your Birth Plan." *Nature Reviews Microbiology* 14 (3): 131.

Ostwald, Madeleine M., et al. 2016. "The Behavioral Regulation of Thirst, Water Collection and Water Storage in Honey Bee Colonies." *Journal of Experimental Biology* 219 (14): 2156–2165.

Palmer, Chris. 2014. "The Necrobiome." *The Scientist* website, February 1, http://www.the-scientist.com/?articles.view/articleNo/38946/title/The-Necrobiome/.

Parton, Ash, et al. 2015. "Alluvial Fan Records from Southeast Arabia Reveal Multiple Windows for Human Dispersal." *Geology* 43 (4): 295–298.

Perkins, Sid. 2013. "Baseball Players Reveal How Humans Evolved to Throw So Well." *Nature News* website, June 26, http://www.nature.com/news/baseball-players-reveal-how-humans-evolved-to-throw-so-well-1.13281.

"Plants Prepackage Beneficial Microbes in Their Seeds." 2014. *ScienceDaily* website, September 29, www.sciencedaily.com/releases/2014/09/140929180055.htm.

Pobiner, Briana. 2016. "Meat-Eating among the Earliest Humans." *American Scientist* 104 (2): 110–117.

Prüfer, Kay, et al. 2012. "The Bonobo Genome Compared with the Chimpanzee and Human Genomes." *Nature* 486(7404): 527–531.

Pugach, Irina. 2013. "Genome-wide Data Substantiate Holocene Gene Flow from India to Australia." *PNAS* 110 (5): 1803–1808.

Quinn, Helen. 2013. "How Ancient Collision Shaped New York Skyline." BBC Science website, June 7, http://www.bbc.com/news/science-environment-22798563.

Quirk, Trevor. 2013. "How a Microbe Chooses among Seven Sexes." *Nature* website, March 27, http://www.nature.com/news/how-a-microbe-chooses-among-seven-sexes-1.12684.

Ratnarajah, Lavenia, et al. 2014. "Bottoms Up: How Whale Poop Helps Feed the Ocean." Science Alert website, August 11, http://www.sciencealert.com/bottoms-up-how-whale-poop-helps-feed-the-ocean.

Raymann, Kasie, et al. 2017. "Unexplored Archaeal Diversity in the Great Ape Gut Microbiome." *mSphere* 2 (1): 1–12.

Reece, Jane B., et al. 2014. *Campbell Biology*. 10th edition. Boston: Pearson.

Reich, David, et al. 2010. "Genetic History of an Archaic Hominin Group from Denisova Cave in Siberia." *Nature* 468 (7327): 1053–1060.

Rensberger, Boyce. 1996. *Life Itself: Exploring the Realm of the Living Cell.* New York: Oxford University Press.

Reservation of Coal and Mineral Rights, 43 U.S.C. §299 (1993).

Ridaura, Vanessa K., et al. 2013. "Gut Microbiota from Twins Discordant for Obesity Modulate Metabolism in Mice." *Science* 341 (6150): 1069–1070.

Roach, Neil T., et al. 2013. "Elastic Energy Storage in the Shoulder and the Evolution of High-Speed Throwing in Homo." *Nature* 498 (7455): 483–487.

Rogers, Alan R. et al. 2004. "Genetic Variation at the MC1R Locus and the Time Since Loss of Human Body Hair." *Current Anthropology* 45 (1): 105–108.

Rogier, Eric W., et al. 2014. "Secretory Antibodies in Breast Milk Promote Long-Term Intestinal Homeostasis by Regulating the Gut Microbiota and Host Gene Expression." *PNAS* 111 (8): 3074–3079.

"Romance Languages." 2017. *Wikipedia: The Free Encyclopedia*, August 12, https://en.wikipedia.org/wiki/Romance_languages.

Rose, C., et al. 2015. "The Characterization of Feces and Urine: A Review of the Literature to Inform Advanced Treatment Technology." *Critical Reviews in Environmental Science and Technology* 45 (17): 1827–1879.

Ross, Robert M. 1978. "Reproductive Behavior of the Anemonefish *Amphiprion melanopus* on Guam." *Copeia* 1978 (1): 103–107.

Rumble, Douglass, et al. 2013. "The Oxygen Isotope Composition of Earth's Oldest Rocks and Evidence of a Terrestrial Magma Ocean." *G3: Geochemistry, Geophysics, Geosystems* 14 (6): 1929–1939.

Sakamoto, M., et al. 2016. "Dinosaurs in Decline Tens of Millions of Years before Their Final Extinction." *PNAS* 113 (18): 5036–5040.

Salvini-Plawen, L. V., et al. 1977. "On the Evolution of Photoreceptors and Eyes." *Evolutionary Biology* 10: 207–263.

Schrag, Daniel P., et al. 2002. "On the Initiation of a Snowball Earth." *Geochemistry, Geophysics, Geosystems* 3 (6): 1–21. http://www.snowballearth.org/pdf/Schrag_2002.pdf.

Scudellari, Megan. 2014. "The Sex Paradox." *The Scientist* website, July 1, http://www.the-scientist.com/?articles.view/articleNo/40333/title/The-Sex-Paradox/.

Seid, Marc A., et al. 2011. "The Allometry of Brain Miniaturization in Ants." *Brain, Behavior and Evolution*. 77 (1): 5–13.

Sender, Ron, et al. 2016. "Revised Estimates for the Number of Human and Bacteria Cells in the Body." *bioRxiv* website, January 6, http://biorxiv.org/content/early/2016/01/06/036103.

"Sex Change in Fish Found Common." 1984. *New York Times* website, December 4, http://www.nytimes.com/1984/12/04/science/sex-change-in-fish-found-common.html.

Singer, Emily. 2014 "In Bees, a Hunt for Roots of Social Behavior." *Quanta Magazine* website, May 6, https://www.quantamagazine.org/20140505-in-bees-a-hunt-for-the-roots-of-social-behavior/.

Singer, Emily. 2016a. "How Neanderthal DNA Helps Humanity." *Quanta Magazine* website, May 26, https://www.quantamagazine.org/20160526-neanderthal-denisovan-dna-modern-humans/.

Singer, Emily. 2016b. "In Newly Created Life-Form, a Major Mystery." *Quanta Magazine* website, March 24, https://www.quantamagazine.org/20160324-in-newly-created-life-form-a-major-mystery/.

Slack, Jonathan. 2014. "A Twist of Fate." *The Scientist* website, March 1, http://www.the-scientist.com/?articles.view/articleNo/39241/title/A-Twist-of-Fate/.

Smolin, Lee. 1979. *The Life of the Cosmos*. New York: Oxford University Press.

Sockol, Michael D., et al. 2007. "Chimpanzee Locomoter Energetics and the Origin of Human Bipedalism." *PNAS* 104 (30): 12265–12269.

Sokol, Joshua. 2017. "A New Blast May Have Forged Cosmic Gold." *Quanta Magazine* website, March 23, https://www.quantamagazine.org/did-neutron-stars-or-supernovas-forge-the-universes-supply-of-gold-20170323.

Spalding, Kirsty L., et al. 2005. "Retrospective Birth Dating of Cells in Humans." *Cell* 122: 133–143.

Strain, Lisa, et al. 1995. "A Human Parthenogenetic Chimaera." *Nature Genetics* 11 (2): 164–169.

Surridge, Christopher. 2003. "Ginkgo Is Living Fossil." *Nature* website, June 19, http://www.nature.com/news/2003/030619/full/news030616-9.html.

Tachon, Gaelle, et al. 2014. "Discordant Sex in Monozygotic XXY/XX Twins: A Case Report." *Human Reproduction* 29 (12): 2814–2820.

Tattersall, Ian. 2015. "Reimagining Humanity." *The Scientist* website, June 1, http://www.the-scientist.com/?articles.view/articleNo/43061/title/Reimagining-Humanity/.

Thomas, David. 2015. "What Other Animals Walk Upright with a Vertical Spine Like Humans?" *Quora* website, January 23, https://www.quora.com/What-other-animals-walk-upright-with-a-vertical-spine-like-humans.

"The Time I Accidentally Married My Cousin." 2013. *A Charleston Accent* blog, September 26, http://acharlestonaccent.com/beautiful-places/2013/9/26/the-time-i-accidentally-married-my-cousin.

Trager, Rebecca. 2016. "Fungi Eat Up Old Batteries and Spit Out Metals." Chemistry World website, August 23, https://www.chemistryworld.com/news/fungi-eat-up-old-batteries-and-spit-out-metals/1017317.article.

Tudge, Colin. 1996. *The Time before History: 5 Million Years of Human Impact*. New York: Scribner.

Twilley, Nicola, et al. 2016. "Why the Calorie Is Broken." Real Clear Science website, January 26, http://www.realclearscience.com/articles/2016/01/26/why_the_calorie_is_broken_109521.html.

United States Department of Agriculture. 2003. *Agriculture Fact Book: 2001–2002*.

United States Department of Agriculture. 2017. "World Agricultural Supply and Demand Estimates." September 12, https://www.usda.gov/oce/commodity/wasde/latest.pdf.

Valley, John W., et al. 2014. "Hadean Age for a Post-Magma-Ocean Zircon Confirmed by Atom-Probe Tomography." *Nature Geoscience* 7: 219–223.

Vernot, Benjamin. 2016. "Excavating Neandertal and Denisovan DNA from the Genomes of Melanesian Individuals." *Science* 352 (6282): 235–239.

Wade, Nicholas. 2016. "Meet Luca, the Ancestor of All Living Things." *New York Times* website, July 25, http://www.nytimes.com/2016/07/26/science/last-universal-ancestor.html.

Walls, Jerry G. N.d. "Breeding Anoles." *Reptiles Magazine* website, http://www.reptilesmagazine.com/Breeding-Lizards/Breeding-Anoles/.

Wang, Huai, et al. 2015. "Evidence That the Origin of Naked Kernels during Maize Domestication Was Caused by a Single Amino Acid Substitution in tga1." *Genetics* 200 (3): 965–974.

Wang, Xu, et al. 2015. "Antibiotic Use and Abuse: A Threat to Mitochondria and Chloroplasts with Impact on Research, Health, and Environment." *BioEssays* 37 (10): 1045–1053.

Weiss, Madeline C., et al. 2016. "The Physiology and Habitat of the Last Universal Common Ancestor." *Nature Microbiology* 1 (16116).

White, Rosalind V. 2002. "Earth's Biggest 'Whodunnit': Unravelling the Clues in the Case of the End-Permian Mass Extinction." *Philosophical Transactions of the Royal Society of London* Series A 360 (1801): 2963–2985.

Wichura, Henry, et al. 2015. "A 17-My-Old Whale Constrains Onset of Uplift and Climate Change in East Africa. *PNAS* 112 (13): 3910–3915.

Wilkinson, Emma. 2008. "Cousin Marriage: Is It a Health Risk?" BBC News website, May 16, http://news.bbc.co.uk/2/hi/health/7404730.stm.

Willbold, Matthias, et al. 2011. "The Tungsten Isotopic Composition of the Earth's Mantle before the Terminal Bombardment." *Nature* 477 (7363): 195–198.

Wrangham, Richard. 2009. *Catching Fire: How Cooking Made Us Human*. New York: Basic.

Yarus, Michael. 2010. *Life from an RNA World: The Ancestor Within*. Cambridge, MA: Harvard University Press.

Yong, Ed. 2016. "Breast-Feeding the Microbiome." *New Yorker Magazine* website, July 22, http://www.newyorker.com/tech/elements/breast-feeding-the-microbiome.

Young, Richard W. 2003. "Evolution of the Human Hand: The Role of Throwing and Clubbing." *Journal of Anatomy* 202 (1): 165–174.

Zeldovich, Lina. 2014. "These Two Guys Studied Their Feces for a Year." *The Atlantic* website, September 3, http://www.theatlantic.com/technology/archive/2014/09/these-two-guys-studied-their-feces-for-a-year/378862/.

Zhang, Yorke, et al. 2017. "A Semisynthetic Organism Engineered for the Stable Expansion of the Genetic Alphabet." *PNAS* 114 (6): 1317–1322.

Zhu, Chen-Tseh, et al. 2003. "Codon Usage Decreases the Error Minimization within the Genetic Code." *Journal of Molecular Evolution* 57 (5): 533–537.

Zihlman, Adrienne L., et al. 2015. "Body Composition in *Pan paniscus* Compared with *Homo sapiens* Has Implications for Changes during Human Evolution." *PNAS* 112 (24): 7466–7471.

Zimmer, Carl. 2013. "And the Genomes Keep Shrinking . . . " *The Loom* website, August 23, http://phenomena.nationalgeographic.com/2013/08/23/and-the-genomes-keep-shrinking/.

Zimmer, Carl. 2016. "Scientists Unveil New 'Tree of Life.'" *New York Times* website, April 11, http://www.nytimes.com/2016/04/12/science/scientists-unveil-new-tree-of-life.html.

Zimmer, Carl. 2017. "Antarctic Ice Reveals Earth's Accelerating Plant Growth." *New York Times* website, April 5, https://www.nytimes.com/2017/04/05/science/carbon-dioxide-plant-growth-antarctic-ice.html.

Zimmer, Marc. 2015. *Illuminating Disease: An Introduction to Green Fluorescent Proteins*. New York: Oxford University Press.

Zmuda, Natalie. 2011. "Bottom's Up! A Look at America's Drinking Habits." *Advertising Age* website, June 27, http://adage.com/article/news/consumers-drink-soft-drinks-water-beer/228422/.

INDEX

Index